Water Treatment Essentials for Boiler Plant Operation

Robert G. Nunn

McGraw-Hill

New York   San Francisco   Washington, D.C.   Auckland   Bogotá
Caracas   Lisbon   London   Madrid   Mexico City   Milan
Montreal   New Delhi   San Juan   Singapore
Sydney   Tokyo   Toronto

Library of Congress Cataloging-in-Publication Data

Nunn, Robert G.
    Water treatment essentials for boiler plant operation / Robert G.
Nunn.
        p.   cm.
    Includes index
    ISBN 0-07-048219-5
    1. Feed-water purification.   2. Steam-boilers—Water-supply
I. Title.
TJ379.N86 1997
621.1′8—dc21   96-47395
                                                                    CIP

McGraw-Hill

A Division of The McGraw-Hill Companies

1 2 3 4 5 6 7 8 9 0   DOC/DOC   9 0 1 0 9 8 7 6

ISBN 0-07-048219-5

*The sponsoring editor for this book was Robert Esposito, the editing
supervisor was Caroline R. Levine, and the production supervisor was
Donald F. Schmidt.*

Printed and bound by R. R. Donnelley & Sons Company.

McGraw-Hill books are available at special quantity discounts to use
as premiums and sales promotions, or for use in corporate training programs. For more information, please write to the Director of Special
Sales, McGraw-Hill, 11 West 19th Street, New York, NY 10011. Or contact your local bookstore.

This book is printed on recycled, acid-free paper containing
a minimum of 50 percent recycled, de-inked fiber.

For Helen
Whom I Love Dearly

Contents

Preface

This book was written to give heating and air conditioning plant operators, and owner-users who are not trained operators, a basic knowledge of the water side of the equipment they are responsible for. The book covers different types of systems—such as cooling towers, hot water boilers, and steam boilers—and their associated equipment.

Management people should find this material helpful in assessing their current practices of operating plant equipment, purchasing chemical feed equipment, purchasing chemicals, as well as hiring operating personnel. It will give them a general knowledge of the equipment for which they are responsible as well.

The book contains simple fundamental explanations of boiler room equipment operation, problems that are encountered when using water with the associated equipment, methods of solving those problems, and basic engineering information. Sample math problems showing necessary computations are included.

Numerous persons and organizations have made significant contributions to this manuscript. I am grateful for their efforts and support and would like to acknowledge their assistance and the help they have given me.

Robert G. Nunn

INTRODUCTION

WHERE DOES WATER COME FROM?

Water covers approximately 70% of the earth's surface and out of all of the water on earth, about 97% is brackish or salty ocean water. Glaciers at the North and South Poles tie up another 2% of the earth's freshwater and the remaining 1% is the water we drink and use in our industries. This usable water is stored in the form of snow-covered mountains, rivers, lakes, and underground rivers and lakes.

One of the advantages we have here on earth is the use of this water. Because water is indestructible, the same water is used over and over again. This process of using water over and over again is called the *hydrological cycle*, (Fig.1). As water moves from the atmosphere to the ground, in the form of snow, rain, or hail, it is used by the earth's plants and animal life and for drinking. Water is also used in industry, and then through a process called *evaporation* and/or *transpiration* the water goes back up into the atmosphere to be used again.

> *Evaporation* is the process of changing moisture or liquid into a vapor.

> *Transpiration* is the process of removing moisture or waste products from the surface of leaves or plants in the form of a water vapor.

Water is one of the oldest of all substances on the planet earth. When the earth was first forming the heat of the earth's mass was so great that chemical combinations could not exist. Then as the gaseous metals and rock-forming elements cooled and changed from a gas to a liquid and then eventually to a solid, the earth's temperatures began to drop low enough to condense the steam vapor into water. Water was one of the first compounds to appear as the earth cooled; then it slowly became a great ocean that covered over 70% of the earth's surface.

The chances are very good that the glass of water you drank this morning was part of that great ocean, formed millions of years ago, for there has been no new water on the planet earth.

HYDROLOGICAL CYCLE

In the hydrological cycle process, water continually is lifted from the surface of the ocean and lakes by the sun's heat. It is then released back to the earth's crust in the form of rain and snow. A great deal of water is also evaporated from lakes, rivers, moist soil, and plants, but most of the evaporated water comes from the surface of the oceans. When temperatures in the upper atmosphere cool, the vapors or gases in the atmosphere combine as pure water and fall to earth in the form of rain, snow, or hail.

WHAT IS WATER ?

Water in its purest form is made up of two gases or elements, *hydrogen* and *oxygen*. One molecule of water is composed of two atoms of hydrogen and one atom of oxygen. The formula for water is

$$H_2O$$

These two gases, hydrogen and oxygen, form an extremely stable compound, which is called *water*.

THREE FORMS OF WATER

1.   Solid state-in the form of ice.

2.   Liquid state-in the form of water.

3.   Gaseous state-in the form of water vapor.

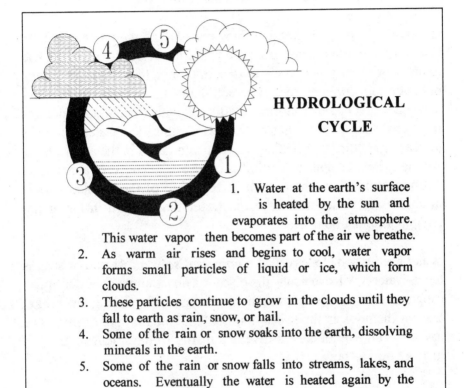

1. Water at the earth's surface
    is heated by the sun and
    evaporates into the atmosphere.
    This water vapor then becomes part of the air we breathe.
2. As warm air rises and begins to cool, water vapor
    forms small particles of liquid or ice, which form
    clouds.
3. These particles continue to grow in the clouds until they
    fall to earth as rain, snow, or hail.
4. Some of the rain or snow soaks into the earth, dissolving
    minerals in the earth.
5. Some of the rain or snow falls into streams, lakes, and
    oceans. Eventually the water is heated again by the
    sun and the cycle is repeated.

FIGURE 1. Hydrological cycle

Water has the ability to change from one of the above noted forms to
another form quite easily, but it always requires energy to do so.
When water changes from one form to another it is given the
following names:

Freezing.   The ability of water (liquid) to change
                  directly into ice (a solid).

Melting.   The ability of ice (a solid) to change
                  directly to water (a liquid).

Evaporation. The ability of water (a liquid) to change directly to a gaseous vapor.

If water were pure H_2O and nothing more, there would be very little need for water treatments. But pure water never occurs in nature and its impurities vary. Nature's closest approach to pure water is rain, but even rain contains certain objectionable impurities. When water condenses and begins to form a cloud, it begins to pick up impurities of many kinds. Starting with these impurities water becomes increasingly "dirtier," picking up dissolved gases in the atmosphere. As the rainwater falls to earth it is quite acidic and very aggressive. An operating plant engineer must be aware of these objectionable impurities to understand why the water that is used in his or her equipment must be treated.

Water itself is sometimes called a *universal solvent*. If iron or steel is left in water it will eventually dissolve. As an example consider a fish caught in the ocean with a steel hook. Suppose the fishing line broke, leaving the hook in the fishes' mouth, *"the one that got away."* The hook would eventually dissolve in a matter of days, leaving the fish just as good as new.

Nature uses different means for storing water on earth: surface waters, groundwater, and ocean waters.

SURFACE WATERS

Surface waters are waters that come from lakes, streams, rivers, and reservoirs above the earth's surface.

When rain falls toward earth, the surfaces of the raindrops pick up other gases and contaminants from the air such as carbon dioxide. Carbon dioxide, when mixed with water, changes into carbonic acid. This causes the rainwater to become slightly acidic. Once the rainwater reaches the surface of the earth this "acidic" water tends to dissolve virtually all minerals it comes into contact with.

Surface waters are waters that travel across the ground into streams, rivers, and lakes. These surface waters reflect seasonal changes due to fluctuations in the amount of rainfall and the melting of the snow and ice at certain times of the year. In general the type of impurities that the water contains depends on the organic materials and minerals the water came into contact with during it's travels across the surface of the ground. The amount of impurities in the water depends on the contact time that the water has had with these impurities.

GROUNDWATER

Groundwater is water that comes from beneath the surface of the earth, such as water wells or underground lakes, rivers, or springs.

As described above when the slightly acidic rainwater falls it travels across the ground and percolates down through the earth, dissolving other minerals that it comes into contact with. As a rule this causes large amounts of dissolved solids to be entrained in the water. The water also picks up fine particles of both organic and inorganic material in its travels. The levels of the dissolved solid matter in the groundwater, which are picked up as the water travels across and down through the earth, depend to a great extent upon the flow rate of the water across the ground and the amount of organic matter growing or decaying in the soil. These contaminants picked up by the water, as we shall see later, may play havoc with metals used in industry.

OCEAN WATERS

Ocean water and brackish waters are used only in a few types of industrial applications due to the high concentrations of salt and other materials in the water. These brackish materials are leached from the earth and finally flow into the ocean as water flows from streams, irrigation wastewater, etc.

As mentioned above the conversion of carbon dioxide to carbonic acid in rainwater will decrease the pH of the water. The decrease of the pH in the water gives the groundwater the ability to dissolve minerals as it passes through the earth's surface. Due to water's ability to dissolve so many materials, water is never in a "pure state in nature."

Contaminants or impurities that the water picks up from the atmosphere and earth can be placed into three broad groups.

DISSOLVED SOLIDS

Many solid materials such as calcium, silica, iron, magnesium, and manganese are dissolved by water. Most dissolved solids, however, originate from contact of water with the earth. Groundwater usually has fairly high levels of dissolved solids, due to the leaching of the minerals from the earth, which is caused by the water percolating through the ground.

SUSPENDED SOLIDS

Suspended solids can also cause turbidity or cloudiness in water. Suspended solids are usually very fine particles such as clay, organic matter, microorganisms, and silt.

Some other materials that will cause turbidity are decaying vegetable matter, industrial wastes, fats, and other effluent. Some vegetable matter may cause the water to take a color, which is a form of turbidity.

GASES

Gases, primarily carbon dioxide and oxygen, are absorbed from the air we breathe by rainwater as it falls to earth. In addition, carbon dioxide and oxygen gases are picked up from decaying matter in the soil as water flows over it.

The dissolved impurities or solids in the water, as outlined above, are of three general types: (1) bicarbonates of calcium and magnesium, (2) sulfates of calcium and magnesium, and (3) sodium salts. Items (1) and (2) were historically called temporary and permanent hardness, respectively, but they are also labeled carbonate and noncarbonate hardness.

(The presence of calcium and magnesium in natural water is the cause of *hardness*, which will produce scaling in boilers, and form a precipitate with soap in a standard soap test solution.) The amount of scale-forming materials may be expressed either in terms of grains of equivalent calcium carbonate, $CaCO_3$, per U.S. gallon, or in parts per million. Parts per million (ppm) is equal to grains per gallon times 17.1.

Waters having bicarbonate hardness may be softened by either sufficient heating or boiling. In this method of purification CO_2 is liberated, and relatively insoluble precipitates of the carbonates $CaCO_3$ and $MgCO_3$ are formed. The equation from the reaction may be expressed for calcium bicarbonates as $Ca(HCO_3)_2$ + heat$\rightarrow$ $CaCO_3$ + $CO_2 + H_2O$. The same general equation applies when $Mg(HCO_3)_2$ is substituted in place of $Ca(HCO_3)_2$ and, then $MgCO_3$ results. The sludge formed may either be discharged with the blowoff water from a boiler or be removed from the feedwater heater. The scales formed by waters having carbonate hardness are softer and more porous than those produced by water having sulfate or noncarbonate hardness. Heating alone will not remove calcium and magnesium sulfate from waters having noncarbonate hardness. Sodium salts in solution, in a boiler, can be removed only by blowing down the boiler when the concentrations become excessive.

Under heat and pressure chlorides and nitrates are prone to break up to form corrosive acids. Sodium carbonate, Na_2CO_3, sometimes appears free in raw waters and quite often in water excessively treated with soda ash. Treatment by the zeolite process produces free sodium carbonate in some waters. When heated, sodium carbonate is very likely to decompose to a hydroxide or caustic soda NaOH, and stressed or strained boiler parts are embrittled by the hydroxide.

Waters are classified, according to their content of scale-forming materials per gallon, as follows: less than 8 grains, very good; 8 to 15, good; 15 to 20, fair; 20 to 30 poor; and more than 30, bad.

WATER TREATMENT

The primary objects of water treatment are twofold: (1) the removal of suspended and soluble solids and (2) the removal of gases. For best results the purification of the water should take place outside of the boiler. A minimum of purification before treatment of any boiler should be a water softener. This is discussed in further detail in later chapters.

Frequent blowing down of a boiler to remove concentrated solutions of salts and precipitated sludge results in a loss of heat in the water and materials discharged. Such loss may be lessened by the operation of continuous blowoff devices that permit the incoming feedwater to be heated by the continuous blowdown water and material discharged so that a part of the heat may be recovered.

Water treatment should reduce to a minimum or absolutely prevent the following undesirable and dangerous conditions: foaming and priming, scale formation, corrosion, and caustic embrittlement.

In order to prevent corrosion the water should be free of oxygen, carbon dioxide, free acids, and materials that may break down to form acids. Care must be taken to prevent excessive concentrations in the boiler of hydrogen ions (H) and hydroxyl ions (OH). The former produce acid reactions and the latter alkaline reactions.

Water treatment should be undertaken only after analysis of the water and recommendations for its treatment have been made by an experienced and competent chemist. No single prescribed process of treatment of boiler compound or chemical treatment is suitable for all waters. Each case must be considered individually.

COOLING TOWERS

The same main characteristics of the water outlined above are also applicable to cooling towers. In response to environmental pressures, the need to conserve water, avoid toxic chemicals, and meet effluent regulations has prompted the development of new and innovative approaches to cooling water treatment. Along with all of these changes, microbiological control methods have changed dramatically in recent years as government regulations and safety considerations have limited the use of both chlorine and nonoxidizing toxic biocides.

Cooling tower makeup water usually comes from one of two sources: surface water or groundwater. Surface water includes river and lake water, and groundwater comes from wells. Other sources include sewage plant effluent and industrial waste streams.

Groundwater is usually harder and contains more alkalinity and total dissolved solids than surface water. On the plus side, groundwater is usually more uniform in quality throughout the year, contains lower turbidity, and is less susceptible to microbiological contamination than surface water.

Surface waters have the desirable feature of being readily available and generally contain lower levels of dissolved solids. However, in some instances it is desirable to use a blend of the two waters.

CHAPTER 1

FUNDAMENTALS

HEAT, A FORM OF ENERGY

Heat is defined as a form of *molecular energy.* Molecules in a body, when heated, increase their activity or movements and when cooled they decrease their movements. This can be demonstrated by putting a pan of water on a stove and heating the water with a burner. As the water heats, you can watch the water begin to move, and the hotter the water gets the faster the activity of the water.

When two different bodies of molecules, one hot and one cold, are brought together the molecules of the hotter body will impart motion to the molecules of the colder body; thus the molecular activity of the hot body begins to decrease and that of the cold body increases until equilibrium of temperature between the two bodies is established. Therefore, heat *is energy* that flows, in response to a temperature difference, between molecules.

This can easily be demonstrated by adding cold water to a pan of very hot water. The molecules in the hot water, which are moving very rapidly, give up some of their energy to the molecules in the cold water. The molecular activity in the hot water then decreases and that in the cold water increases until temperature equilibrium is established.

TEMPERATURE

Temperature is a measurement of the intensity of the heat as distinguished from the quantity of heat; temperature is *the thermal state of a body considered with reference to its ability to transmit heat to other bodies.* The temperature of a body may be obtained from a measurement of the intensity of the radiant energy it emits.

There are two scales used to measure temperature. One is *Fahrenheit,* and the other is *Centigrade.* The difference between the two measurements is the scale or the *means of measurement.*

A Fahrenheit thermometer is divided into 180 equal parts, and a Centigrade thermometer is divided into 100 equal parts. If both thermometers were inserted into freezing water or melting ice at standard atmospheric pressure, the Centigrade thermometer would indicate 0° freezing, while the Fahrenheit thermometer would indicate 32° freezing. For the boiling point of water, the Centigrade thermometer would indicate 100° C and the Fahrenheit thermometer would indicate 212° F.

Wet-Bulb Thermometer

A wet-bulb thermometer is one whose bulb is encased within a wetted wick. The primary basis for using the wet-bulb thermometer is for the thermal design of any evaporative-type cooling tower. The operator of a cooling tower as a rule will never be required to take this measurement, but should be aware of how it works.

Wet-bulb temperatures are measured by causing air to move across a thermometer whose bulb is *properly shielded* and encased in a wetted muslin "sock." This is accomplished by swinging the wetted-shielded thermometer through the air in a circle. As the air moves across the wetted bulb, moisture evaporates from the sock or wick, and the sensible heat in the thermometer is transferred from the mercury in to the wick, cooling the mercury and causing an equilibrium to be reached at what is called the *Wet-Bulb temperature. Sensible heat* is the difference between the dry-bulb heat and the wet-bulb heat. This is sometimes referred to as "the approach to wet-bulb temperature."

TRANSMISSION AND MEASUREMENT OF HEAT

There are three methods by which heat may be transmitted or conveyed from one place to another, namely *CONDUCTION, CONVECTION,* and *RADIATION.*

Conduction

Conduction is the transfer of heat from one part of a material to another part of the same material or to a another body with which it is in physical contact, but without any displacement of the particles of the bodies. As an example, the heat from a soldering iron will conduct heat to the solder when they come into contact with each other, but the soldering iron will not transfer or displace any of its metal to the solder.

Convection

Convection is the transfer of heat from a fluid (liquid or gas) by currents. Convection can further be defined as *free* or *forced* convection.

Free Convection

When cold water flows over a heated surface, the heat is transferred to the water, heating the water. Heated water has a lighter density and rises while cooler water has a heavier density and descends to displace the heated water. The difference in temperature between the hot and cold water causes currents that move the water. If the water did not circulate, eventually equal temperatures would result and the transfer of heat by convection would stop.

Forced Convection

This operates in the same way as free convection, except that an external mechanical force is used to move the liquid or gas. As an example, a hot-water heating system uses an external pump to force the water over or through a boiler and then through the heat distribution coils.

Radiation

All bodies give off heat in the form of radiant energy. Radiant energy can be compared to the light that is emitted from a lamp. Radiation falling upon a body is absorbed by that body, either wholly or in part. When two bodies at unequal temperatures are placed within an enclosure, there is a continuous interchange of radiant energy between them. The hotter body radiates more energy than it absorbs, and the colder body absorbs more energy than it radiates. After the bodies reach an equilibrium in temperature the process still continues, each body radiating and absorbing energy.

ENERGY

Energy is the capacity for doing work and the ability to overcome resistance; energy can exist in many forms, such as heat energy, mechanical energy, electrical energy, and chemical energy. Energy transferred from one body to another body by conduction, convection, or radiation can be measured by a common unit of energy called the *British thermal unit* (Btu).

The British thermal unit is the quantity of heat required to raise the temperature of 1 lb of pure water 1° F at normal atmospheric pressure. The Btu is not an exclusive unit of heat. Any other form of energy may also be expressed in Btu's, such as foot pounds, gram calories, horsepower hours, and kilowatt hours.

1 Btu = 788.16 foot pounds   (Ft · lb)
1 Btu = 252.0 gram calories   (g · cal)
1 Btu = 0.0003927 horsepower hours (hp · h)
1 Btu = 0.0002928 kilowatt hours   (kw · h)

WORK

When a force acting on a body causes that body to move, the force is said to do *work*. In general, *work is the overcoming of a resistance over a distance.* Thus, if F represents the magnitude of the force and D the distance through which the force acts,

$$Work = F \cdot D$$

If F in the equation above is in pounds and D is in feet, the work W will be in footpounds. If D were in inches, the work W would be in inchpounds.

POWER

Power is the rate at which work is performed. Any working agent is said to be developing *one horsepower* when it does 33,000 ft · lb of work in one minute.

$$Power = work \cdot time$$

Power Equivalents

1 horsepower = 746.00 watts (W, units of electrical power)
1 horsepower = 42.44 Btu / min
1 horsepower = 33,000.00 (ft · lb) / min
1 boiler horsepower hour = 34.5 lb of water evaporated per hour

1 boiler horsepower hour = 4.0 gal of water evaporated
per hour
1 boiler horsepower hour = 33,520.0 Btu/hour

SPECIFIC GRAVITY OF LIQUIDS

The specific gravity indicates how much a certain volume of a liquid
weighs compared with an equal volume of water. *By definition, the
specific gravity of water is 1.* Sulfuric acid has a specific gravity
number of 1.84 because it is 1.84 times as heavy as water.

PRESSURE

The earth's atmosphere exerts variable pressures on the earth,
depending upon the altitude and weather conditions. Observation has
shown that the average atmospheric pressure at sea level is 14.7 lb/
in^2 .

For all practical applications gages indicating pressures above that of
the atmosphere are called *pressure gages.* Those showing pressures
less than atmospheric pressure are known as *vacuum gages..*

Pressure gages are usually made with a Bourdon tube, which consists
of an elastic metal tube of oval cross section bent into an arc.
Attached to the Bourdon tube is a needle. As the internal pressure in
the Bourdon tube changes, the needle will move, indicating the
changes on the face of the gage. This type of gage is used for
determining the pressure of closed systems, such as steam, water, and
air, and it is called *gage pressure.*

Gage pressure plus the atmospheric pressure is called *absolute
pressure.* Plant operators normally are only interested in gage
pressure. It is the gage pressure that they are monitoring as they walk
through the plant. Engineers when designing a plant might be
interested in absolute pressure, but even then they look at gage
pressure.

ACIDS, BASES, AND pH

The term pH, despite its widespread use, is seldom understood. By scientific definition, *pH* is the logarithm of the reciprocal of the hydrogen ion concentration.

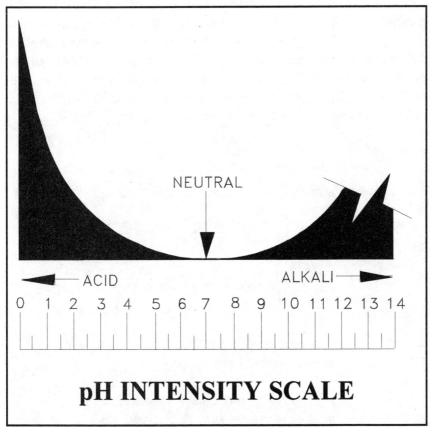

pH INTENSITY SCALE

FIGURE 1.1 pH is a term used to denote the relative intensity of acidity or alkalinity of a given solution. The comparative intensity of acidity or alkalinity may be determined from a pH intensity scale. pH is further defined as the logarithm of the reciprocal of the hydrogen ion concentration of a solution.

When reading a pH scale as shown in Fig. 1.1, the pH numbers 5 and

6 are 10 times more intense than the numbers between 6 and 7, and
the pH numbers between 4 and 5 are 100 times more intense than the
pH numbers between 5 and 6, etc. For all practical purposes, pH is a
measurement of the acidity or alkalinity of a solution. The numbers
0 to 14 express pH values denoting the intensity of an acid or a base
depending on which side of neutral they are. A pH of 7.0 (at 25° C)
indicates neutrality. The human eye has a pH of 7.0. A pH over 7.0
indicates alkaline solutions, while a pH below 7.0 indicates acid
solutions. The closer the pH is to 7.0, the weaker the acid or alkali is.
The farther away from the pH of 7.0, the stronger the acid or the
alkali. A solution having a pH reading of 6.5 is only very slightly acid,
while a solution having a pH of 5.0 is more intensely acidic, and a
solution having a pH of 1.5 is extremely acidic.

SAFETY

When working around acids or caustics, *safety masks and gloves* are
a must.

Things to Remember

1.   Always have someone with you when working
      with acids or caustics.

2.   Wear protective clothing, such as a safety
      mask, gloves, and an apron.

3.   Always relieve any static pressure in an
      acid or a caustic system that you are
      working with before opening any
      valves. Be sure that all of the pumps
      that might operate and that might
      supply acid or caustic to the system are
      turned off. If at all possible, drain any
      acid lines into a suitable container
      before dismantling them.

4.   Always use rubber gloves when handling parts that have acid or caustic on them. Rinse the parts you are working with, using clean water, and then dry them before handling.

5.   Always pour acid into water *slowly.* Pour a little acid at a time, because the chemical reaction may cause considerable heat in the container into which you are pouring the acid.

6.   Be very careful when washing down the area around an acid container. Do not direct the water stream from a hose into or onto the container. If water is poured into an acid container, it could cause a fire or an explosion because of the chemical reaction of the water being added to the acid.

7.   Never allow a low-pH material to come into contact with a high-pH material. This would be similar to rocket fuel and could cause a fire or an explosion.

8.   The same rules, as outlined above, also apply when handling a high-pH or very caustic material.

9.   *Be careful:* know what you are going to do before you do it.

PARTS PER MILLION AND RATIOS

Do not be alarmed when using equations. Take each word of an equation, like the word *"million"* and give it a letter, like *"M"* and put that letter into the equation until you have the formula you want.

Parts

A *part* is a portion of a whole. One part in 10 parts = one of 10 equal parts. The same thing is true with larger numbers, such as one part in one million parts = one part of a million.

An equation for the above would look like this:

$$\text{Parts} = P$$
$$\text{Million} = M$$

$$\frac{1P}{M} = \frac{1P}{1{,}000{,}000 \ parts}$$

Reading the above you would say "One part per million parts." The equation above can also be written as

$$\frac{1 \ part}{1{,}000{,}000 \ parts} = 0.000001 \ of \ a \ part$$

Ratios

Ratios are the relationship of two quantities. A ratio of 3 to 5 is expressed as:

$$3{:}5 \ \ \text{or} \ \ 3/5$$

PROPORTIONS

Proportions are a relationship of equivalence between two ratios, for example, the equation:

$$\frac{a}{b} = \frac{c}{d}$$

The equation above is made up of two ratios. One ratio states that a is related to b in the same way that the other ratio c is related to d. The above equation can also be written as

$$a{:}b :: c{:}d$$

Equivalent ratios are said to be in *proportion.*

In the proportion $a{:}b :: c{:}d$, a is called the first term; b is called the second term; c is called the third term; d is called the fourth term.

The first and fourth terms are called the *extremes* Both a and d are on the extreme ends of the formula $(\boldsymbol{a}{:}b{::}c{:}\boldsymbol{d})$.

The second and third terms are called the *means.* (middle). Both b and c are in the middle of the formula $(a{:}\boldsymbol{b}{::}\boldsymbol{c}{:}d)$.

Rule 1

For all true proportions the product of the means equals the product of the extremes.

Example

For the ratio   3:4 :: 6:8   or   3/4 = 6/8

    The means $(4 \cdot 6) = 24$
    The extremes $(3 \cdot 8) = 24$

The means and extremes *are in balance*. Because of this rule, if any of the above terms is an unknown term, it can be solved for by using one of the following formulas.

Ratio Equations

$$a/c = c/? \qquad d = a/(b \cdot c) \qquad (1.1)$$
$$a/b = ?/d \qquad c = (a \cdot d)/b \qquad (1.2)$$
$$a/? = c/d \qquad b = c/(a \cdot d) \qquad (1.3)$$
$$?/b = c/d \qquad a = (c \cdot b)/d \qquad (1.4)$$

Rule 2

The units in ratio equations must always remain the same, for example, pounds, ounces, tons, gallons etc.

Example of How Ratios are Used

Problem A

If 3 oz of chemical is required to treat 1000 gal of water in a cooling tower, how much chemical would be required to treat 100,000 gal of water?

Solution

     Chemical = C

     C : 1000 gal :: 100,000 gal : ?

     C · 100,000 gal / 1,000 gal = 300 oz of chemical

Problem B

A chemical supplier says that the chemical product *x y z* requires 1 pt to treat 1000 gal of makeup water in a steam boiler and the chemical costs $2.80 a gal.

1.   How much of the chemical supplier's
      xyz chemical will be required to treat
      a 300 hp boiler, operating at 70% load
      with 10% condensate return, for 30
      days, estimating 10% blowdown?

2.   What will be the chemical cost, for 30
      days, using the chemical supplier's
      chemical *xyz* ?

Computation

1.   *1 boiler hp h = 4 gal of water
      evaporated per hour*:

      300 hp · 4 gal of water evaporated per h =
      1,200 gal of water per hour.

      24 h per day · 30 days · 1200 gal per h
      = 864,000 gal of water evaporated every 30 days.

2.   *70% loading on the boiler*:

      0.7 · 864,000 gal = 604,800 gal

      604,800 gal of makeup water required.

3.   **Estimating 10% blowdown:** Boiler blowdown water
      is lost to both drain and evaporation, so the amount of
      blowdown water must be added back into the
      amountof makeup water required.

$$\frac{604,800 \ gal}{1 - 0.1} = 672,000 \ gal \ makeup \ water$$

4.   ***Estimating 10% condensate return***: Condensate is returned to the makeup water, so we can subtract this amount of water from the 672,000 gal.

10% of 672,000 gal of makeup is

0.1 · 672,000 gal of makeup = 67,200 gal

672,000.0 - 67,200.0 = 604,800.0 gal of makeup required

Rounding off the 604,800 gal of makeup required we find that 605,000 gal of makeup is required.

5.   *Using the ratio formula* [Eq.(1.2)]:

1 pt : 1000 gal = ? pt : 605,000 gal

$$\frac{1\ pt}{1000\ gal} = \frac{?\ pt}{605,000\ gal}$$

$$1\ Pt \cdot \frac{605,000\ gal}{1000\ gal} = 605\ pt$$

Solution

1.   605.0 pt of chemical are required. (There are 16 pt in 1 gal.)

$$\frac{605\ pt \cdot gal}{16\ pt} = 38\ gal$$

38 gal of x y z chemical would be required to treat a 300 hp boiler operating at 70% load with 10% condensate return.

2. At $2.08 per gallon the operating cost would be 38 gal · $2.08 per gallon, or $79.00 for one month.

CHAPTER 2

COOLING SYSTEMS

BACKGROUND

Evaporation is *the change of a substance from a liquid to a gaseous state.*

When molecules of a liquid, such as water, are heated they begin to move faster. When the molecules begin to move fast enough to break away from the surface of the water they are in effect, evaporated into the atmosphere.

One of the rules for evaporation is that when a liquid evaporates, it must first absorb heat and then the evaporation takes that heat away. Therefore when a molecule of water is evaporated from water, that molecule must first absorb heat from the surrounding water and then take that heat away, reducing the temperature of the remaining water.

As an example of one of the most ancient forms of natural cooling, moisten the back of your hand with some water or alcohol. Then move your hand rapidly back and forth in the air. The heat from your hand causes a molecule of alcohol or water to begin to move very rapidly, and as the molecule breaks the surface of the alcohol or water it evaporates into the atmosphere, taking the heat from your hand with it. Remember as the molecule of alcohol or water leaves the back of your hand and evaporates, it must first remove heat from your hand.

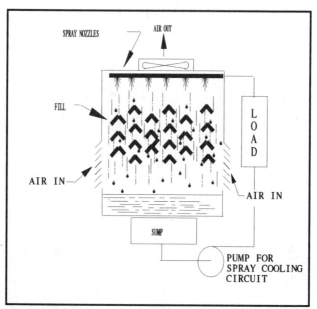

FIGURE 2.1 Spray-filled counterflow tower.

When your skin temperature gives up heat in this manner, the skin temperature goes down, cooling the back of your hand.

In prehistoric days, perspiring humans depended upon natural breezes to accelerate the evaporative cooling process on their bodies. Later,

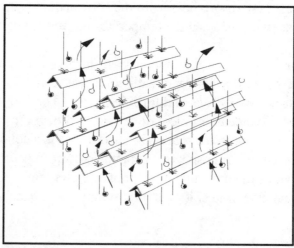

FIGURE 2.2 Splash-type fill.

in the distant past, slaves swept broad palm leaves to create an artificial breeze, and unknowingly the basic concept of a cooling tower was founded.

All cooling towers, other than air-cooled towers, operate on the principle of evaporative cooling. When water evaporates, it must first absorb heat.

This is a universal chemical principle. Tower water that is made to evaporate must first absorb heat from the surrounding water, thereby cooling it.

Evaporation as a means of cooling water is utilized to its fullest extent in cooling towers. Cooling towers are designed to expose the maximum transient water surface to the maximum flow of air for the longest possible period of time.

The *spray-filled counterflow tower* shown in, Fig. 2.1 attempts to accomplish this basic chemical principle by first spraying the water into fine droplets and then containing those droplets so that they fall through a mechanically induced moving stream of air.

Fill was then added to the tower to impede the progress of the falling water. Figure 2.2 shows a splash-type fill that is placed in the horizontal area of the tower, below the sprays and above the air inlet level in staggered rows. These splash bars retard the falling water droplets and increase the surface area exposed to the upward flowing airstream, thereby promoting the process of evaporation.

WET-BULB TEMPERATURE

The primary basis for thermal design of any evaporative-type cooling tower is the *wet-bulb temperature* of the air entering the tower. (*See Chap.1, section titled "Wet-Bulb Thermometer."*)

Selection of the *wet-bulb* design temperature for a cooling tower must be made on the basis of the conditions existing at the site, being proposed for the cooling tower. The wet-bulb temperature should be that which will result in the optimum cold water temperature at or near the time of peak load demand for the cooling tower.

CYCLES OF CONCENTRATION

Every time water is evaporated from a cooling tower system and that same amount of water is then replaced with new makeup water, the dissolved solids in the water increase by the same amount that was in the water when the tower was originally filled.

For example, let us assume that in the original filling of a cooling tower, the water had a silica content of 20 ppm. Also assume that the water in the tower was then evaporated (without any bleedoff). If the evaporated water is replaced with fresh makeup water having the same silica content of 20 ppm, the silica content was added to the first 20 ppm or doubled from the original 20 to 40 ppm / gal. At that point in time the water in the tower had two cycles of concentration. Every time water is evaporated from a tower system and replaced with an equal amount of makeup water, the cycles of concentration increase by one, provided that bleedoff water from the tower is not taken into account. Controlling the number of cycles in a cooling tower or boiler is *very important.*

Both the amount of chemicals being used in a cooling tower and the amount of water used are determined by the concentrations of water in the tower.

MAKEUP TOWER WATER

For example, if the above-noted tower system held 3000 gal of water, there was no makeup, and the water was evaporating at a rate of 10gal / min, theoretically the tower would stop working after about 5 h, because there would be no water left in the tower to evaporate.

In order to keep a cooling tower operating, the tower must have makeup water coming into it at the same rate that the tower water is being evaporated. This is performed by adding makeup water to the tower. To do this, a float valve is mounted in the tower basin that senses the water level in the tower. When the water level falls below the set level of the float valve, the float valve opens, allowing makeup water to enter the tower. This inflow of water brings the water level in the tower basin back up to the set lever of the float valve. This maintains a balance between the evaporated water and the makeup water in the tower.

BLEED (BLOWDOWN)

If the water condition above were allowed to continue without bleed, the tower cycles would build up to five, six, seven, ten, or even more cycles of concentration. It becomes obvious that if the evaporation process in a cooling tower continues and the dissolved solids continue to accumulate in the tower system indefinitely, only one result will occur. The solids in the circulating tower water would increase tremendously, jeopardizing not only the cooling tower but the heat exchanger and all other associated components. The water finally would be so saturated with solids that they would start to precipitate out of solution and begin to be deposited in the piping, on the cooling tower fill, in heat exchanger, etc. This scaling of the system can become very costly as the weight of the solids, which are like rock, could break down the tower. Any tower water analysis would call for more chemical to be added, which would just be wasted, and the efficiency of the tower would drop to zero, producing no cooling whatsoever. Operating under these conditions, the tower would eventually have to be shut down, cleaned, or replaced in its entirety, which is a very costly procedure.

To prevent this buildup from happening, the highly concentrated solids (hardness) must be removed from the system through bleed. The lost bleed water has to be replaced with an equal amount of clean makeup water, which is comparatively less concentrated with solids. This indispensable process is called *bleeding* or *bleedoff*, or in the case of a steam boiler, *blowdown*.

The amount of bleed required can be expressed as follows:

$$Bleed = \frac{Evaporation\ rate}{Cycles\ of\ concentration\ -1}$$

Figure 2.3 shows a graph of cycles of concentration plotted against the gallons of water used by a 500-ton tower over a period of 24 h. It is easy to see from the graph that a cooling tower uses much more water as the cycles of concentration are reduced. This means that more water is going to waste as the cycles of concentration become less.

Example

The difference shown on the graph between 2 cycles and 7 cycles of concentration is 2000 gal of water used per day. A savings of 2000 gal over a period of a year would amount to over 730,000 gal of water.

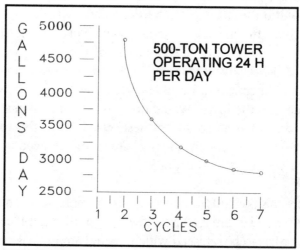

FIGURE 2.3 Tower water consumption.

It also should be remembered that the less makeup water used in a cooling tower, the less chemical treatment will be required to treat the tower, so that there is a considerable cost savings in the use of chemical when the cycles are increased.

It has been determined that an amount of water roughly equal to 2 gal of water per ton per hour will be evaporated into the atmosphere. So to prevent the eventual shutting down and scaling of the cooling system, 2 gal of water per ton per hour should be bled to waste.

COOLING TOWERS

Cooling tower systems are designed and manufactured in many different configurations for many different applications of cooling. Although plant operators may never see or use the different types of towers, they should have some knowledge of their configuration and operation and how to take care of them.

Their are two basic types of evaporative cooling tower cooling systems used to cool liquids. The first type of tower is called *the spray-filled counterflow type*. This tower involves direct contact between the heated water and the atmosphere (Fig. 2.1)

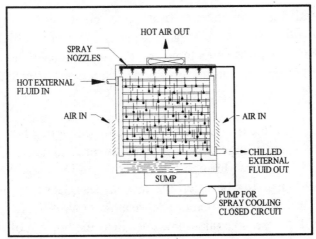

FIGURE 2.4   Fluid cooler.

The second type is called a *fluid cooler* (see Fig. 2.4). This device does not expose the process water directly to the atmosphere. The heated material to be cooled is transferred to the cooling tower through tubing, which passes under the tower spray water, thereby transferring the source heat from the tubing directly to the air through the evaporation process. The material in the closed tubing is never exposed to the atmosphere.

Fluid Coolers

The indirect-contact tower (*closed-circuit fluid cooler*) contains two separate fluid circuits: (1) the external circuit in which water is exposed to the atmosphere as it cascades over the tubes of a coil bundle and (2) an internal circuit in which the fluid to be cooled circulates inside of the tubes in the coil bundle. In operation, heat flows from the internal fluid circuit, through the tube walls of the coil, and to the external water circuit, which is cooled by evaporation. As the internal fluid circuit never makes contact with the atmosphere, this unit can be used to cool fluids other than water and prevent contamination of the primary cooling circuit with airborne dirt and impurities.

Natural Draft Tower

The natural draft tower, sometimes referred to as the *hyperbolic natural draft tower,* is extremely expensive but very efficient. Its thermal performance is predictable and the tower is used in very large plants such as electrical atomic energy plants, where they have to cool large quantities of water.

The operation of the natural draft tower is quite unique (see Fig. 2.5). The tower is like a huge chimney; some towers stand as high as 500 ft in height. Airflow through the tower is produced by the density differential that exists between the heated, less dense air inside the stack and the relatively cool, more dense ambient air outside the stack. Hot air tends to rise, so colder air comes in at the bottom of the stack

to replace the hot air. Water sprays, located in the bottom of the
towers, release water over the hot-water heat exchanger, causing
evaporation to take place as the high updraft currents take the heat
away.

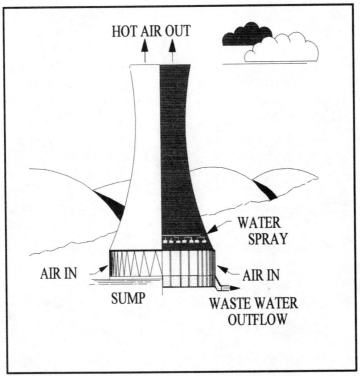

FIGURE 2.5 Counterflow natural draft tower.

CHAPTER 3

MEASUREMENTS

BACKGROUND

As mentioned in the introduction, *evaporation* is the change of a substance to a gaseous state. When molecules of a liquid such as water are heated, they begin to move faster and faster. The molecules begin to move so fast that they break the surface of the water and in effect are evaporated to the atmosphere. One of the rules of evaporation is that when a liquid evaporates it must first absorb heat from that liquid and then take that heat away as the liquid dissipates into the atmosphere. So when a molecule of water is evaporated into the atmosphere, that molecule must first absorb the heat from the surrounding water, hence reducing the temperature of that water.

Evaporation of water is utilized to its fullest extent in cooling towers. Cooling towers are designed to expose the maximum transient water surface to the maximum flow of air for the longest possible period of time.

Evaporation of water in a cooling tower is demonstrated in Fig. 2.1, which shows a mechanically induced airstream flowing from the sides of the tower up through and out the top of the tower. The water flowing down through the tower is first broken up into fine spray droplets by the spray nozzles in the tower. This fine spray is then induced into the upward moving stream of air where the droplets are held for the longest possible period of time. To better cool the water, "fill" has been added to the tower in the form of splash bars (Fig. 2.2). The splash-type fill is placed in the horizontal area of the tower below the water sprays and above the air inlet. These splash bars retard the falling water even more and at the same time increase the wetted surface area that the tower water is exposed to, thereby promoting the process of evaporation.

FACTORS AFFECTING COOLING TOWER PERFORMANCE

The atmosphere from which a cooling tower draws its supply of air incorporates infinitely variable psychometric properties, and the tower reacts thermally and physically with each of those properties. The cooling tower is designed to accelerate the air as it passes it through a maze of structure and fill. It heats the air, expands the air, saturates the air with moisture, scrubs the air, and then it compresses the air and during all of this, the air responds to all of the thermal and aerodynamic effects that such treatment can produce. Finally, the cooling tower returns the used-up stream of air to the atmosphere. In most cases, atmospheric winds do not reintroduce the air back into the tower.

WET-BULB TEMPERATURE

Selection of the design for the *wet-bulb temperature* for a cooling tower must be made on the basis of the conditions existing at the site for the proposed cooling tower, and the wet-bulb temperature should be that which will result in the optimum cold water temperature at or near the time of the peak load demand of the tower. The lowest theoretical temperature for cooling water, due to evaporation, is its *dewpoint* (*the wet-bulb temperature*). However, no cooling tower is 100 % efficient, so the cooling tower is normally designed to cool to within 8° to 15° of the wet-bulb temperature. This is called the *approach to wet-bulb temperature.*

In most cases, 1 % of the water passing over a tower is evaporated so as to cool the remaining 99 % of the water. For example, let us look at a cooling tower operating at a 10° temperature drop across the tower (known as delta T), and a recirculating rate of 1000 gal / min.

The tower will evaporate 10 gal / min to atmosphere, cooling 990 gal of the remaining water. The delta T is measured by taking the temperature of the water entering the tower and subtracting the temperature of the water leaving the tower. This is a very important measurement, which is used in many of the calculations regarding cooling towers.

MEASUREMENT OF CYCLES OF CONCENTRATION

Chlorides. Along with calcium, iron, magnesium, and other compounds found in water, there is a very helpful compound called *chloride.* Chlorides are found in almost all natural waters. Table salt is a chloride.

Chloride solution concentrations or chloride salts in water will not decrease in the same way that concentrations of calcium will. Calcium solution under high concentrations precipitates out of solution, while chlorides keep building on themselves. Because of this, under normal operating conditions or at very high concentrations and temperatures, the measurement of chloride is a very valuable tool in measuring cycles of concentration.

The test for chlorides in the water is very simple and reliable and is used as a measuring tool to determine the number of times that water is evaporated and replaced in cooling towers and steam boilers, as well as for other applications. (*You should be able to purchase a chloride test kit from your water treatment representative.*)

FOR EVERY RULE THERE IS AN EXCEPTION

If additional chlorides such as table salt were to be added before a test to a cooling tower system's concentrated water, the chloride test would be invalid. The same would be true if chlorine used to kill algae in a cooling tower were added to a cooling tower water system during a chloride test.

TESTING FOR CYCLES

To determine the cycles of concentration in a cooling tower it is necessary to take two samples of water: one sample of the incoming makeup water and one of the tower water. Perform a chloride test on each of the samples and then divide the incoming makeup water chloride number into the number of chlorides of the tower water. The result is the number of cycles of concentration in the tower.

Example

$$Cycles = \frac{Tower\ Water\ Chlorides}{Makeup\ Water\ Chlorides}$$

This would also be true for testing the cycles in a steam boiler.

The second choice for determining cycles of concentration when a chloride test method cannot be used is *electrical conductivity.*

Electrical Conductivity

Electrical conductivity, referred to as *total dissolved solids* (TDS), is the measurement of the conductivity of the water. For example, pure water will not conduct electricity. But if salt, (*chloride)*, is added to the same pure water, electricity will flow through the water.

Conductivity is measured in mho at a specified temperature in C. The unit for resistance is the ohm Resistance is the reciprocal of conductance, so the natural unit for conductance is the *mho* (the reverse spelling of ohm). The conductivity test is a measurement of the conductivity of electricity and because of the small amount of conductance that is being measured, the units are given as *micro-mho* (μmho) per centimeter. (1 μmho is 1/1,000,000 of a mho.)

Conductivity testers are used for measuring conductivity of water, and a water treatment sales person should be able to sell you one.

Conductance in water also can be used as an indication of the dissolved solids in the water, but as described above, it cannot be specifically related because some dissolved substances (silica, for example) contribute little or nothing to conductivity. Should some of the other solids precipitate out of that solution, conductivity would not be a reliable measurement.

The same procedure, taking two samples of water for the chloride test, one from the incoming makeup water to the tower and one from the tower water, is used to determine cycles of concentration. This time the conductance of the different waters is used to determine cycles of concentration using a conductivity tester.

Remember that the TDS in the water do not always give a true reading of the cycles of concentration because some of the dissolved solids in a cooling tower, such as calcium bicarbonate, may have exceeded their solubility limit and precipitated out of solution, laying down excess calcium bicarbonate as scale on the pipe walls of the tower system and other equipment.

The operator performing the TDS test may get a reading of four cycles of concentration; however, the tower system may be actually operating at ten cycles because six of the cycles have precipitated out of the water and onto the tower walls, piping, and other parts of the cooling systems. Therefore conductivity is not always the best way to measure cycles of concentration.

TOWER CALCULATIONS

The following calculations will be helpful in determining the tonnage, evaporation rate, bleed rate, makeup, amount of chemical required, etc., for cooling towers.

Tonnage

Tonnage is the name of measurement that is used for refrigerating capacity, and the unit is called a *ton* of refrigeration. This equals the amount of heat required to melt 1 ton of ice (2000 lb) in 24 h and corresponds to a heat removal of 12,000 Btu / h or 200 Btu / min.

Cooling towers are also rated in tons. A practical and easy way to calculate the tonnage for a cooling tower is as follows:

$$Tonnage = \frac{gal/\text{min} \ Circulated \ tower \ water}{3}$$

or

$$Circulated \ tower \ water \ (gal \ / \ min) \ = Tonnage \ \cdot \ 3$$

Evaporation

When water evaporates from a cooling tower it also removes heat from the surrounding water. The temperature of the water leaving a cooling tower is less than the temperature of the entering water. Depending on the geographical location of the cooling tower and the wet-bulb design temperature for that location, the tower will have different temperatures of water entering and leaving. This temperature is called the delta ($\triangle$) T of the cooling tower, *the temperature drop across the tower.*

The evaporation rate of a cooling tower is measured in gallons of water evaporated per minute. The tower evaporation rate in gal / min is calculated as follows:

$$[Circulated \ tower \ water \ (gal \ / \ min)] \ \cdot \ 0.001 \ \cdot \ \Delta \ T$$

Wind and Splash Loss

Wind outside a tower sometimes blows the water being evaporated inside the tower out through the sides of the tower. This water, *wind loss*, is lost to the evaporation process of the tower. At the same time, water falling down through the tower causes splashing when it hits bottom and sometimes splashes outside of the cooling tower. This is called *splash loss*. There are also other losses of water in the process water as well, such as water in the tower leaking through the cracks in wooden towers. This would also be considered *wind and splash loss*.

The amount of wind and splash loss in small towers is relatively small and is not taken into consideration as a rule.

The wind or splash rate is given as follows:

$$gal \ / \ min \ = \ 0.0002 \ \cdot \ [Circulated \ tower \ water \ (gal \ / \ min)]$$

Bleed (Blowdown)

The bleed rate is given as follows:

$$Bleed \ =$$

$$\frac{Evaporation \ (gal \ / \ min)}{(Number \ of \ cycles \ - \ 1) \ - \ [Wind \ (splash) \ loss(gal \ / \ min)]}$$

Makeup Water

Makeup water is the total amount of water required to replace the water lost due to the evaporation, bleed, and wind loss. The makeup water is given as follows:

Evaporation (gal / min) + Wind (splash) loss (gal / min) +

bleed (gal / min) = Makeup water (gal / min)

HOW TO USE THE EQUATIONS NOTED ABOVE

PROBLEM

What is the total makeup water required?

Assumptions

Tower	300 tons
ΔT of Tower	10° F
Required cycles	6 cycles of concentration

Computations

1. Tower circulation rate:

    300 tons · 3 = 900 gal / min

2. gal / min Evaporation rate:

    900 gal / min (tower circulation rate) · 0.001
    · 10 gal / min = 9 gal / min

3. Wind and splash loss rate:

    9 gal / min · 0.0002 = 0.0018 gal / min

4. Bleed (blowdown rate):

    $$\frac{9 \ gal \ / \ min}{6 \ cycles \ - \ 1} - 0.0018 \ gal \ / \ min = 1.798 \ gal \ / \ min$$

5.   Total makeup water required:

+   9.0000 gal / min (evaporation loss)
+   .0018 gal / min (wind and splash loss)
+   1.7980 gal / min (bleed or blowdown loss)
=   10.7998 gal / min makeup required

Solution

10.7998 gal / min makeup required

STEAM BOILER MEASUREMENTS

Most of the measurements for a steam boiler are chemical analyses of the different waters taken from different places, such as makeup water, condensate return water, feedwater, boiler water, blowdown water, and condensate water. It is important to know where to take these samples of water so that the measurements are meaningful.

Where to take samples

Treated water. This would be either city water or water that has passed through a treatment process such as water softening or demineralization.

Makeup water. In most cases boiler makeup water would be the water feeding into a feedwater tank, a deaerator, or in some cases directly into the boiler.

Condensate return water. This is a collection of condensate return water coming from steam traps throughout the operation. This is not condensate. It is condensate return water. Sometimes condensate return lines are corroded and the water traveling through them will pick up contaminants, such as oxygen and groundwater.

Condensate water. The condensate water sample should be taken directly from the discharge of the steam trap. Care should be taken if there is no valve on the trap from which to take a water sample. If the trap is not working properly and the plug at the end of the trap is removed, live steam under pressure could escape and cause the operator to be burned. Sometimes it is necessary to take condensate samples from all of the condensate traps in order to locate a trap that is not operating properly or to find a leak in the system that is letting in oxygen.

Feedwater. This is a combination of condensate return and makeup water. This sample should be taken just before the feedwater pump.

Boiler water. Sometimes it is difficult to get a sample of boiler water, especially when the design engineer did not make provisions for taking a water sample out of the boiler. Boiler water should be cooled before it comes into contact with the atmosphere, so that steam does not evaporate from the sample, which would make it more concentrated. Sometimes a sample of boiler water can be taken through the small valves located on the feedwater control valve. The operator must have pliers to turn the valve as it is very hot and always under boiler pressure.

Blowdown water. This water is also boiler water, and care should be taken in collecting a sample as it is also very hot and under boiler pressure.

Taking samples of water from the locations noted above will allow the operator to determine some of the following measurements: cycles of concentration in the boiler, percentage of condensate return to the boiler, percentage of makeup to the boiler, condition of the makeup water, amount of chemicals in the boiler, amount of chemicals in condensate, acidity of condensate and percentage of blowdown.

When working with boiler water, one should always convert gallons of water to pounds of water. Most calculations used to determine horsepower, etc., are in pounds.

As an example, let us assume you wanted to know how much water a 300 hp boiler uses so you can determine how much chemical should be used. Looking in Appendix A, 1 boiler hp evaporates 34.5 lb of water per hour.

$$\frac{300 \ boiler \ hp \ \cdot \ 34.5 \ lb \ evaporated.}{h} = 10,350 \ lb \ of \ water$$

If the boiler was receiving 85% condensate return, the amount of makeup water would be as follows:

$$100\% - 85\% = 15\%$$

$$\frac{15\% \ \cdot \ 10,350 \ lb}{h} = \frac{1,552.5 \ lb}{h}$$

Converted to gallons of water this would be

$$\frac{1,552.5 \ lb \ \cdot \ gal}{h \ \cdot \ 8.33 \ lb} = \frac{186.31 \ gal}{h}$$

CHAPTER 4

TOWER PROBLEMS

SCALE

Scale is an inorganic deposit resulting from crystal deposition of supersaturated solutions. *(See the section titled "Cycles of Concentration" Chap. 2.)*

Types of Scale

Calcium Carbonate ($CaCO_3$): the breakdown of calcium bicarbonate.

Calcium Sulfate: Calcium and sulfate.

Calcium Phosphate: High amounts of calcium and orthophosphate. Usually caused from too much water treatment (Sometimes referred to as *phosphate hideout* in boilers).

Silica: High concentrations of silica.

Iron oxide: Compound caused by lack of corrosion control or naturally occurring iron being oxidized.

Iron phosphate: Compound caused by poor film formation of the phosphate inhibitor.

Others: Manganese oxide, high levels of manganese

oxide. Magnesium silicate, high levels of silica and
magnesium with a high pH. Magnesium carbonate,
high levels of magnesium with a high pH and high
CO_2.

Questions about controlling scale deposits should be referred to your
water treatment supplier as there are so many different considerations
and ways of controlling scale. The type of scale and its origin, for
example, must be considered. Some scales can be prevented by
softening the feedwater and removing the calcium and magnesium,
some by using acid, some by using different chemicals such as
phosphates, polymers, etc.

MICROBIOLOGICAL ORGANISMS

Microbiological organisms are usually expressed in plate counts where
bacterial colonies are incubated in a volume of the sample and
counted. The reported value is usually expressed as most probable
number of colonies per 100 milliliters of sample volume. Other
organics are expressed as carbon extractibles (CCE), chemical oxygen
demand (COD), and biochemical oxygen demand (BOD).

Relative humidity in the tower has a significant effect on the control
of airborne infection. At 50% relative humidity, the mortality rate of
certain organisms is at its highest. The mortality rate decreases both
above and below this value. High humidities can support the growth
of pathogenic or allergenic organisms. To deter the propagation and
the spread of these detrimental microorganisms, periodic cleaning of
the tower should be performed. Biocides, which are used to kill
microorganisms, should also be used.

Because of the inherent conditions under which cooling towers
operate, they are very good breeding grounds for algae, slime-forming
bacteria, fungi, and other microorganisms. Because cooling towers
suck in air and then wash that air with water droplets, towers are
continually being bombarded with fresh supplies of organisms that are
present in the air passing through the tower.

These microorganisms tend to seek out the environment best suited to their growth. Algae blooms on the distribution deck, because of the sunlight. Bacteria and fungi tend to collect in dark areas, where they can settle and adhere to the metal surfaces.

Deposit-Producing Organisms

1. *Fungi*. Molds and yeasts: characterized as relatively large organisms, usually no color, and act as filter for mud and silt and other organisms.

2. *Algae*. Mainly blue or blue-green. Characterized as large stringy filamentous organisms always with some color; require sunlight; act as filter for mud and silt and other organisms.

3. *Bacteria*. Non-spore-forming and spore-forming slime formers. Characterized as very small gelatinous, mucous, sometimes colored organisms; act to bind all types of suspended material together, particularly mud and silt, corrosion products, and other organisms.

Corrosion-Causing Bacteria

1. *Sulfide-producing bacteria* (also known as sulfate-reducing bacteria). Bacteria that produce hydrogen sulfide (from sulfate or other sulfur source); attack metals, primarily mild metals and stainless steels; reduce chromates and precipitate zinc salts; destroy chlorine; live in the absence of oxygen (anaerobic); generally appear under slime bacteria or other deposits.

2. *Sulfuric-acid-producing bacteria*. Organisms that have the ability to convert hydrogen sulfide to sulfur compounds and then to sulfuric acid; the low pH corrodes most metals, primarily mild steel. Generally

live with other organisms.

3. *Nitrifying bacteria.*   Organisms that have the ability
    to convert ammonia and ammonia compounds to nitric
    acid; the low pH corrodes most metals, primarily mild
    steel; difficult to test for. Live with all types of other
    organisms.

4. *Iron bacteria.* Organisms   that   convert   soluble
    (ferrous) iron to insoluble (ferric) iron and utilize the
    energy of the conversion. Do not produce corrosion
    but cause it to continue more rapidly by removing the
    soluble iron from solution; readily found where water
    only passes over a heat exchanger once (one pass
    system) or over a potable system;   often found in
    shallow wells.

Decay-Causing Organisms

1. *Fungi.* Organisms that usually produce internal rot of
    wood, converting the cellulose wood product to a
    food source; results in loss of wood strength. Live in
    warm, moist, conditions and cause decay in internals
    of wood.

2. *Bacteria.* Organisms that utilize a cellulose and lignin
    as a food source, causing a softening and loss of
    strength on the surfaces of wood.

Legionnaire's Disease Bacterium

The biggest cooling tower problem is that they are located on
rooftops or in out-of-the-way places. They could be characterized as
"out of sight, out of mind." Towers seem to be the last to be
maintained or looked at by maintenance or management people.

Towers should be inspected at least three times a year; more often is preferable. As the air is pulled into a tower, the water washes the dirt, cottonwood seed, bird, dung, etc. into the tower. Eventually the dirt and mud become so thick that the tower is actually circulating mud throughout the system.

In 1976, over 200 people developed pulmonary infections while attending the Philadelphia American Legion Convention. The causative bacterial agent was subsequently isolated and identified as Legionnaire's disease bacterium (LDB).

Pilot studies found that LDB was demonstrable in all test towers and that LDB appears to be dependent upon biomass and other organisms being present in the system. The study found that LDB was also demonstrable in air conditioning (low temperature) and large industrial type (high temperature) towers.

Organic biocides indicated a relative ineffectiveness to control LDB. High dosages of chlorine (1.5 to 1.7ppm free chlorine) for up to 96 h followed by daily 1-h shock treatment (0.5 to 0.7 ppm) of chlorine controlled LDB populations in some cooling towers.

For best control of LDB in cooling towers, keep the tower basins as clean as possible by washing them down frequently.

BIOCIDES

Microbiocides function to inhibit the growth of microorganisms. Some alter the permeability of cell walls, thereby interfering with vital life processes, while others destroy protein essential to life support. Biological treatment is broken down into two broad categories: oxidizing or nonoxidizing. Consult your water treatment supplier as to the best microbiocides to use for your particular case.

Feeding of Biocides

When biocides are introduced into a cooling system the dissolved solid in the tower immediately increase. If the bleed is being monitored by a dissolved solids controller it will immediately call for bleeding water from the tower. This in turn will bleed both water and chemicals from the tower water. For that reason the controller should be deactivated for at least two to three hours when feeding biocide so the biocides can do their work and not be removed from the tower.

Chemical dosages for biocides are usually expressed in either ounces, pounds, or pints per 1000 gallons of water in the system. Therefore the plant operator should know the volume of water in the tower system to be treated so that the proper amount of chemical can be added to the system.

Keeping the cooling tower clean and free of mud and sediment is the best solution for preventing the growth of *fungi, algae,* and *bacteria,* and for the prevention of *corrosion.* A cooling tower is an air washer, and it scrubs all of the incoming air with water. It collects insects, fly ash, plant and seed fibers, animal hair, clothing fibers, cigarette butts, paper, leaves, grease, pollen, and more. All of this settles in the tower basin or the low point of the system. This accumulation of dirt and debris is a natural habitat for growing micro-organisms, and can be dangerous. A dirty tower basin robs the tower system of the treatment chemicals and acts as a breeding ground for microorganisms. Cooling tower basins should always be kept clean, and whenever noticeable debris accumulates the tower should be manually cleaned out.

SOLID SEPARATORS AND SAND FILTERS

Solid separators and sand filters are mechanical methods for the removal of debris entrained in cooling water systems on a regular basis. This is done usually through the use of an automatic programmable motor-controlled ball valve. One advantage of the solid separator is that the pressure drop through the separator is quite small and consumes very little energy. Another advantage is that when the separator dumps dirt to drain, it only dumps a few gallons of water along with the dirt. Solid separators are an economical water-saving device

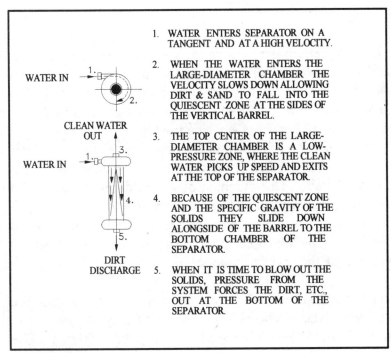

1. WATER ENTERS SEPARATOR ON A TANGENT AND AT A HIGH VELOCITY.

2. WHEN THE WATER ENTERS THE LARGE-DIAMETER CHAMBER THE VELOCITY SLOWS DOWN ALLOWING DIRT & SAND TO FALL INTO THE QUIESCENT ZONE AT THE SIDES OF THE VERTICAL BARREL.

3. THE TOP CENTER OF THE LARGE-DIAMETER CHAMBER IS A LOW-PRESSURE ZONE, WHERE THE CLEAN WATER PICKS UP SPEED AND EXITS AT THE TOP OF THE SEPARATOR.

4. BECAUSE OF THE QUIESCENT ZONE AND THE SPECIFIC GRAVITY OF THE SOLIDS THEY SLIDE DOWN ALONGSIDE OF THE BARREL TO THE BOTTOM CHAMBER OF THE SEPARATOR.

5. WHEN IT IS TIME TO BLOW OUT THE SOLIDS, PRESSURE FROM THE SYSTEM FORCES THE DIRT, ETC., OUT AT THE BOTTOM OF THE SEPARATOR.

FIGURE 4.1   Tower Separator.

Separators

Solid separators use centrifugal force to separate the debris from the particle weight and buoyancy such as sand and dirt. These materials are easily removed because of their specific gravity. The solid separator does not remove everything but it is inexpensive and does a good job of keeping the tower basin clean. Constantly recirculating the cooling tower water through a solids separator will remove approximately 98% of the solids in the tower. These separators are usually installed in a direct line or a bypass line from the cooling tower circulating pump, so that all or a percentage of the dirt-laden cooling tower water always passes through the solid separator where the dirt and sand are removed (see Fig. 4.1). The clean water then passes back to the tower. Dirt collected by the separator should be dumped to drain on a regular basis. This is done usually through the use of an automatic programmable motor-controlled ball valve.

Sand Filters

Sand filters preform the same service as solid separators except that a sand filter forces a percentage of the tower water through *filter media*. To force the tower water through the filter and to backwash the filter requires a small pump that is constantly operating. Sand filters will remove just about anything from the tower water, but it can only filter a portion of the water flowing over the tower. When the sand filter is backwashed it requires a large quantity water that is dumped to drain. This dumping of water, during backwash, does several things to the chemical balance of the tower water. It removes chemicals from the tower water system, disrupts the total dissolved solids (TDS) balance of the tower water, and puts the filter out of service during the backwash mode.

To feed chemicals properly to the tower when using a sand filter, a water meter type of chemical feed should always be used (see the section entitled *"Water Meter Pulse Feed"*). In addition, the overall first cost of the sand filter is considerably more expensive than a separator and will also cost more to operate.

BLEED VALVE

Bleeding water from a cooling tower is one of the most important functions of a cooling tower and consideration should be given to the *bleed valve* and the dirt strainer located just upstream from the bleed valve.

The standard practice, and one that is recommended by the valve companies, is to put a strainer ahead of the solenoid valve, so that the small orifices in the valve will be protected from being plugged up and causing the valve to fail. Most solenoid bleed valves, when they fail, will fail in the open position.

Dirt strainers are usually installed in the piping line ahead of a solenoid valve. These strainers are very seldom cleaned and when a strainer fills up with dirt the flow of water through the strainer is stopped. In the case of a cooling tower bleed system, this plugged strainer will stop any bleeding from the tower. Because most solenoid bleed valves fail in the open position, it is best to remove the strainer, or the strainer screen from the strainer so that the strainer will not be blocked with dirt and prevent the tower from bleeding. This will also allow the tower to bleed in the event of a failed solenoid bleed valve.

Another problem with bleed valves is that if a bleed valve is designed for zero back pressure, meaning that it does not require any water pressure to operate the valve, there must still be a minimum of 3 to 4 lb/in^2 head pressure upstream of the valve. This is so that there will be enough water pressure above the valve to push the water through the valve to drain. Should the bleed valve be designed for 10 lb/in^2 back pressure there should be enough water pressure (a minimum of 10 lb/in^2) upstream of the valve to open it.

The best location for the bleed valve is downstream from the condenser or heat exchanger. The reason for this is that when the tower system bleeds water it should be bleeding warm water. This will reduce the cooling load on the tower. On large towers this can be very important.

Example

What would be the savings of dumping the bleed right after the condenser when the circulating water is at its hottest, rather than allowing the hot water to go through the tower, using the energy of the tower to cool the hot water and then dumping the water through the bleed valve to waste ?

Assumptions

Tower, 100 % load	500 tons
ΔT (temperature drop across the tower)	10° F
Cycles of concentration	5

Evaporation and bleed

1. $\dfrac{500 \text{ tons } \cdot 3 \text{ gal/min}}{\text{ton}} \cdot 0.01 = 15 \text{ gal/min evaporation}$

2. $\dfrac{15 \text{ gal/min}}{5 \text{ cycles } -1} = 3.75 \text{ gal/min bleed}$

3. $\dfrac{3.75 \text{ gal/min} \cdot 60 \text{ min} \cdot 1 \text{ lb} \cdot 1 \text{ Btu} \cdot 1 \text{ Ton} \cdot 10^{o}F}{\text{min} \cdot 1 \text{ H} \cdot 1 \text{ gal} \cdot 12{,}000 \text{ Btu}}$

$$= \dfrac{1.56 \text{ ton}}{H}$$

4. $\dfrac{1.56 \text{ ton}}{H} \cdot 24 \text{ H} = 37.44 \text{ tons}$

5.　　Result:　A savings of 37.44 tons of refrigeration in 24 h.

TOWER CONTROLLERS

Feed and Bleed

This type of tower controller senses the total dissolved solids in the tower water. It has a set point so that when the TDS of the tower water reach this set point the controller will close a relay, sending a voltage to the bleed valve and also to the chemical feed pump. Thus, the controller will control the feeding of chemical and control the bleeding of the tower water at the same time.

Caution should be taken that there is no other source of bleed; otherwise this system will not operate properly.

Water Meter Pulse Feed

This type of tower controller operates in two modes using a water meter in the water makeup line to the tower.

1. When the TDS of the tower water reach the set point, the controller will close a relay, sending an electric signal to the bleed valve to bleed water from the tower.

2. A water meter installed in the makeup water line to the cooling tower pulses the controller after a set amount of gallons have passed through the meter. This pulse then starts an electrical timer that controls the time during which the chemical feed pump will operate.

The above are two basic methods of controlling the chemical feeding and bleeding of a cooling tower. Extra functions can be added to the controllers, such as solid state electronics, digital readout versus linear readout, temperature-compensated sensors for TDS, flow switches, pH control, and microprocessor controls. The operating principle is the same for all of them.

Microprocessors

Microprocessors are now being built to pass on tower information to a central office where all of the tower operating functions, such as TDS, pH, bleed, chemical tank levels, and chemical feed, are recorded. If one of the tower functions should get out of balance, an alarm notifies the plant operator what that problem is.

CHAPTER 5

TOWER TREATMENTS

Dramatic changes in water treatment practices and technology have been made over the past 50 years. Even now these changes are being updated continuously due to new technology and environmental government pressures.

In recent years water conservation has emerged as an environmental problem to be considered along with pollution control when planning water treatment programs. Water treatment companies are striving for the ultimate in conservation and quality of products so there are no scale, no corrosion, and no biological problems. Some cooling tower treatments are approaching zero blowdown and some chemical manufacturers are beginning to use more and more chemicals that are nontoxic.

Understanding water treatments for cooling towers begins with an understanding of the characteristics of the cooling system itself. (*See Chapter 2 section titled "Background."*) Basic water-cooling systems consist of

> Closed system recirculating towers
> Open system recirculating towers

Regardless of the size or type of cooling system, tower system water problems are very similar in that they are dependent on the makeup water used, water velocity, water temperature, pH, the materials of tower construction, the type of heat exchangers, the type and amount of blowdown, environmental restrictions, and the type of water treatment used. All of these problems fall into three main categories:

Deposit formations
Corrosion
Biological deposition and corrosion

The rate at which scale and corrosion problems occur is not a constant, and it complicates the water treatment program for any system. A specific chemical treatment program depends on the quality of makeup water available and the end quality of cooling water needed to minimize corrosion and deposits.

DEPOSIT PROBLEMS

Scale deposit problems occur when water-soluble minerals and inorganic materials reach their limit of solubility and precipitate out onto the piping and heat-transfer surfaces of the mechanical equipment.

There are three basic methods for preventing formation of calcium scale in cooling-water systems:

1.   *Removing calcium hardness or scaling minerals from the water prior to its use.* Scale-forming constituents can be removed by the lime-soda softening process, ion exchange, or reverse osmosis. When water is treated in this manner there is a tendency for the water to become more aggressive, *corrosive,* requiring closer control of corrosion inhibitors in the system.

2.   *Keeping scale-forming constituents in solution.* The two polymeric organics most commonly used are polyacrylates and modified polyacrylates. The unmodified polyacrylates seem to be the most popular due to their lower cost and are considered most effective. For good calcium scale control a molecular weight of 1000 has proven to be acceptable, and a dosage rate of 3 to 5 ppm seems to keep calcium scale from forming. These new chemicals, even under severe scaling conditions, when properly used can effectively prevent calcium carbonate

deposition under alkaline conditions. Polyacrylates perform well at high temperatures and are not degraded easily. They are stable and soluble at high and low pH levels and act as good dispersants for suspended material. Concentrations of over 100 ppm may cause a calcium polyacrylate scale.

Disadvantages of polyacrylates are that they react with strong cationic materials, such as biocides, which remove the polyacrylates from solution and reduce their scale control capabilities.

Advantages of polyacrylates are that they are nontoxic and present no problems from an environmental standpoint.

3. *Allowing the impurity in the water to precipitate out as a removable sludge, rather than as a hard deposit.* This method allows scale to form but uses chemicals to distort the resulting crystal structure, changing the scale to a non-adherent sludge, which tends to drop out in the cooling tower. Crystal modification of scale is fast becoming a popular method of scale control due to the fact that much higher cycles of concentration can be maintained with solubilizing chemicals. It is also more effective at higher temperatures and results in more cost savings. There are two classes of crystal modifiers: Polymaleic acids and sulfonated polystyrenes.

Both modifiers are synthetically produced and are classified as water-soluble polymers. They are effective against calcium scale as well as calcium sulfate and calcium phosphate. In severe scaling environments they are dosed between 0.5 and 5 ppm but the normal range for cooling waters is 0.5 to 2 ppm. Both classes of modifiers are compatible with chlorine but can be deactivated with strong biocides. Since crystal modifiers produce sludge, treated cooling water will appear turbid. Sludge must be removed through bleed and/or side-stream filtration.

FOULING PROBLEMS

Fouling in cooling towers is due to the presence of insoluble suspended solids in the circulated water. They include matter that was originally dissolved in the water but that has precipitated out under chemical reactions. Foulants can occur naturally or they can be produced artificially. Typically natural foulants are silt, bacteria, and mud. Artificial foulants are chemical metallic corrosion products and wood cooling tower fibers that have been introduced into the system.

Both scale and fouling work together on heat-exchanger surfaces, contributing to heat loss and reduction of the heat-exchanger efficiency.

Fouling control in cooling-water systems has become more important than ever. This is because cooling systems are being operated over longer periods of time between clearings and at higher water temperatures and heat-transfer rates.

Sludge fluidizers, surfactants, and wetting agents constitute accepted foulant-control agents. Dispersants function by breaking up the foulants into smaller particles and keeping them suspended in the cooling water, enabling foulant removal from the system via blowdown or filtration.

Synthetic polymers represent a major improvement over natural polymers for several reasons: they can be made with any specific molecular weight, are not easily degraded by biological organisms, do not react with chlorine or iron salts, and most important cost less for the same performance.

A combination of mechanical and chemical prevention plus chemical-cleanup methods can be used with great success in keeping towers free of scale and corrosion. Sidestream filtration or separation can be very effective in removing substantial concentrations of foulants from the entire recirculating-water loop. (*See Chapter 4, section entitled "Solid Separators and Sand Filters."*)

Good housekeeping is the best way of controlling fouling problems.

BIOCIDES

Oxidizing Microbiocides *Chlorine and bromine* are both oxidizing microbiocides. They are less expensive than nonoxidizing microbiocides but are dangerous to handle, are corrosive to metal, and cause delignification of wood. Bromine is more effective at higher phs and chlorine is less effective.

Chlorine is the most familiar and effective industrial oxidizing biocide. Chlorine has long been used as a disinfectant for domestic water supplies and to remove tastes and odors from water. When chlorine gas is introduced to water, it hydrolyzes to form hydrochloric and hypochlorous acids. It is the hypochlorous acid that determines the effectiveness of the biocide. This acid is an extremely powerful oxidizing agent, diffusing easily through the cell walls of microorganisms, reacting with the cytoplasm to produce a chemically stable nitrogen-chlorine bond with the cell proteins. It tends to lose its effectiveness at higher pH.

Increasing concern for the environmental effect of escaping chlorine gas, the residual effect of bleach (liquid chlorine) or sodium hypochlorite has led to regulatory limitations on its use. Sodium hypochlorite is an alkaline water solution of chlorine and can simplify feeding while minimizing the dangers of handling pressurized poisonous gas. Calcium hypochlorite is a powder, commonly known as HTH. Calcium in the dry chlorine will also contribute hardness to the cooling tower waters. One advantage of chlorine is that it is less expensive than many of the nonoxidizing agents.

Bromine is the latest material offered as a cooling tower biocide. It is superior to chlorine as a biocide and is a very strong oxidizing biocide used for microbial control of cooling water because of its ability to detoxify rapidly. A continuous residual of 0.1 ppm bromine is equivalent to 0.5 ppm chlorine. It is more effective at higher pH, has a lower vapor pressure than chlorine, and is six times as soluble in water, making it less subject to vaporization loss in a cooling tower.

From an environmental standpoint the hydrolysis or breakdown of bromine is significant. Its decomposition is accelerated by both pH and temperature. Half-life of the molecule depends strongly on the pH, dropping rapidly from 2 hr to 2.5 sec as pH varies from 8 to 11.0. Because of this, bromine is one of the better biocides to use for environmental purposes. The disadvantage of bromine is that it is harder to handle and harder to feed. Bromine burns are painful and slow to heal. Both chlorine and bromine tablets are strong oxidants and should be handled with gloves to prevent burns.

Caution:

When using chlorine or bromine, a chemical feeder should always be used. If bromine or chlorine tablets are just thrown into the cooling tower, they will settle to the bottom of the tower, creating a very concentrated, strong corrosive spot under the tablet on the floor of the tower. This can eventually destroy metal and cause a tower to leak.

NONOXIDIZING MICROBIOCIDES

Nonoxidizing microbiocides are toxicants that have proven to be effective against a broad spectrum of microorganisms. There are a variety of nonoxidizing toxicants that are currently in use.

Ozone and Ultraviolet Light

Ozone is an unstable pale blue gas formed by an electrical discharge of air. Manufacturing ozone involves the passage of oxygen (usually contained in air) between electrodes across which a high-voltage alternating potential is maintained. The two electrodes are powered from a step-up transformer. Ozone exists as O_3 and is slightly soluble in water. As an unstable gas ozone does not remain as a residual in water. It will dissipate or revert back to oxygen within 15 to 20 minutes after formation. The advantage of using ozone is that it can be produced on site without keeping a chemical inventory of hazardous compounds. Ozone does not react with ammonia and is not pH dependent, although a pH of 6.0 to 7.0 is needed for optimum ozonation. Ozone requires 5 minutes contact time to effect the same kill as 30 minutes for chlorine. The cost of ozone equipment is about four times that of chlorine feed equipment, and as a rule, some conventional treatment will also be required.

Ultraviolet Light

Ultraviolet light operates somewhat the same as ozone. Experimental work indicates that disinfection with ultraviolet radiation is similar to that by chlorine. Both ozone and ultraviolet light are considered impractical for applications to anything but small volumes of water.

Dibromo-nitrilo-propionamide (DBNPA)

The DBNPA molecule is an extremely potent, broad-spectrum microbiocide, and its mode of action is very fast. It permeates the microbe's cell membrane, ties up certain protein groups, and stops the oxidation-reduction process, and then the organism dies. Decomposition of DBNPA is accelerated by both increased pH and temperature. Detoxification can be accomplished by raising the temperature and/or the pH.

Organo-Tin Compounds

These compounds are effective against algae and wood-rotting organisms. Because they remain nonionized in solution, they readily penetrate an organism's cell wall and invade the cytoplasm, where they form complexes with amino and carboxyl acid protein groups. The amino acids are thus modified and rendered useless for normal life-sustaining reactions.

Quaternary Ammonium Salts

These salts are most effective against bacteria in the higher pH ranges and work better in towers that are clean. Overfeeding can cause excessive foaming in the tower. These cationic, surface-active chemicals were originally substituted for nitrogen compounds, which were generally most effective against algae and bacteria in the alkaline pH ranges. The electrostatic bonds, created by long chains of carbon, initiate stresses in the wall of the organism, leading to cell lysis and death. The quaternary ammonium salts tend to lose their activity in systems that are heavily fouled with dirt, and if overfed, they can produce extensive foaming.

Organo-Sulfur

This product is highly effective against fungi and slime-forming bacteria, particularly sulfate-reducing bacteria, and is water-soluble and easy to use. Organo-sulfur compounds have been used widely for biocidal purposes in paper and food production, but they are also available for use in cooling-water systems. Through a chemical process removal of a ferric ion from the cytochrome stops a transfer of energy and causes immediate cell death.

Methylene bi-Thiocyanate

This is one of the most effective microbiocides for inhibiting algae, fungi, and bacteria growth, most notably the sulfate-reducing bacteria. It is not effective in heavily fouled systems with dissolved ferric iron. It is pH sensitive in ranges above 7.5, and it is not recommended for use in high alkaline waters. Since it is a competitive-type microbiocide, its killing action is to block the transfer of electrons in the microorganism, resulting in cell death. The compound is usually formulated in combination with a dispersant to increase its solubility in water systems and enhance its penetration of algae and bacterial slime layers.

In some situations, nonoxidizing toxicants can prove to be more effective than their oxidizing counterparts.

BIOLOGICAL DEPOSITION AND CORROSION

Biological problems are not always caused by microscopic organisms. These problems can be caused by large plant and animal organisms such as weeds, dead rats, floating debris, and other animal organisms. Mechanical screening can remove most of this type of problem but the smaller organisms pass right through and still remain a problem.

Microbiological problems such as algae, slime, fungi, and other microorganisms breed easily in cooling tower environments because they are continually being washed into the tower water by the incoming air, which passes through the tower. Gelatinous slimes produced by many microorganisms can trap these sediments, thus encouraging fouling scale, and they also can cause corrosion. Visual inspection of biological deposition has its drawback because if you can see deposition, the condition has already gone too far. To maintain an effective microbe-control program, it is necessary to measure the numbers both of the total microorganisms present in the system and of each specific type. In this way, the success or lack of success of the program can be determined and corrective action taken as required.

Microbiocides function to inhibit the growth of microorganisms. Some biocides alter the permeability of cell walls in organisms, thereby interfering with vital life processes, while others destroy protein essential to life support of the microorganism.

Selection of a microbiocide involves several factors. It must first be effective in inhibiting practically all microbial activity and it must be an economical treatment program. Frequency of the treatments has great bearing on the cost of the program, and consideration should also be given to the environmental pollution or the disposal of the biocide due to its toxicity. Your water treatment company should be consulted to determine the best microbiocide to use for your particular operation.

Microbiocides are usually slug-fed to the tower system in order to incur rapid effective population reductions from which the organisms cannot easily recover. Retention time in the system is very important in determining the effectiveness of the program.

CORROSION INHIBITORS

Anodic inhibitors

Anodic chemicals are chemicals that work to interfere with the anodic reaction, and perform well by isolating the corrosion cell. (See Chapter 7, Fig. 7.2.) While they reduce the anodic areas available on metal surfaces, they are rarely able to eliminate all potential corrosive regions, and pitting is likely. Because of this anodic inhibitors are called *dangerous* inhibitors.

Chromate had probably been the best and most often used anodic inhibitor, but it is harmful to the environment. Chromate at high levels will protect mild steel, copper, and aluminum alloys, and it is often used with many other inhibitors to improve its effectiveness. Because any blowdown with chromate is very toxic and harmful to the environment, as a rule it is not used anymore.

Nitrites are very suitable for closed systems. They are designed to provide a film or insulated area to reduce the anodic area available on metal surfaces. To be effective, a minimum of 600 ppm of nitrite is required to maintain the film. Nitrites are very unsuitable for open recirculating systems since they oxidize and provide a nutrient supply for organisms. Moreover, any blowdown is very toxic to animal life with nitrites harmful to the environment.

Orthophosphate is now the most commonly used anodic inhibitor. To be effective, a minimum of 40 to 60 ppm of the phosphates is required for film formation and 20 ppm or more is required for maintenance of the film. Periodic monitoring of the treatment levels, corrosion characteristics and rates, and deposit formation can ensure the protection of equipment.

Cathodic Inhibitors

These chemicals function by creating a dense corrosion-resistant film on the metal surface. Cathodic inhibitors are generally considered to be *safe* inhibitors because they do not promote pitting attack.

Polyphosphate forms a thin film or a slight scale containing calcium, iron, and phosphate over the cathodic surface. It requires a minimum of 100 ppm calcium hardness to properly form the protective film, and tends to form a scale if calcium hardness exceeds 600 ppm. By reverting to orthophosphate however, they also furnish their own anodic inhibitor. This reversion product is a potential scale former with calcium and a foulant with iron. These chemicals have a tendency to build on themselves, promoting fouling in waters with flow rates less than 1 ft/s. With higher flow rates of 15 ft/s the protective films are easily removed, leaving the metals unprotected. Polyphosphates do not protect copper or aluminum but will actually attack them. Polyphosphates are also nutrients for algae growth in cooling towers and when discharged they can contribute to algal blooms in lagoons.

Zinc is an effective cathodic inhibitor but it only works for mild steel. At high pH levels, it also can contribute to fouling.

Molybdates are used as cathodic inhibitors for mild steel. Applied alone they require fairly high concentrations. For closed systems, levels of 100 to 200 ppm are needed to inhibit corrosion. For open recirculating systems, about 10 to 20 ppm is required. Usually molybdates are too costly to be used alone and are economical only for closed systems.

Polysilicates perform well in protecting steel, copper, and aluminum from corrosion. They do not perform well at pH values below 7.5, but can be used up to pH 9. Polysilicates should not be used in waters containing more than 150 ppm of natural silica. Their protective film is difficult to form and easily removed. They have no biological effects and are considered nontoxic.

The control of scale deposits, fouling, and microbiological organisms should be referred to your water treatment supplier, as there are so many different considerations and ways of controlling these problems.

PREDICTABLE SCALING INDEX

This is a method for calculating the pH of the cooling tower water to provide a way to determine if the tower water is in a scaling condition. The main objective of using these indices is to adjust the cooling-water chemistry to a nonscaling condition. It should be cautioned that this process is to test for the predictability of scale being formed at the most probable location where scale is expected. It does not test for corrosion. The determination of scale being formed is a calculation of the pH of saturation (pH_S) for calcium carbonate, that is, the solubility constant of $CaCO_3$ under a specific water condition. This requires knowledge of the following values for the system's cooling water.

> Calcium hardness (as $CaCO_3$)
> Total alkalinity (as $CaCO_3$)
> Total dissolved solids
> Maximum temperature where scale is expected

When taking temperature readings for these calculations, always remember to measure the temperature at the hottest point in the system.

The method used for predicting scaling is rather complex and takes time to perform, but if it is practiced on a regular basis, it is quite helpful. To simplify the operation and for the purpose of keeping records, a format has been included which will help in the operation.

Record on the predictable *tower scaling profile* (Fig. 5.1) chart the measured TDS, temperature, calcium hardness and total alkalinity of the tower water to be tested. Using the Tables 5.1 to 5.5, find the factor number for each of the above and record them on the chart. Add the factor totals in each column then subtract the second total from the first total. Multiply this total by 2, which is the pH Saturated. Subtract the equilibrium pH value found in (Table 5.6) from the Saturated pH. This will give the *Ryznar scaling index number*, which will tell what the predictable scaling severity is for this calculation. This number can be checked against Table 5.1, which will then give the condition for that Ryznar index number. Write the condition for this number on the PREDICTABLE TOWER SCALING PROFILE under "Scaling Severity is".

TABLE 5.1

Scaling Severity Keyed to Indices.

RSI*	Condition
3.0	Extremely severe.
4.0	Very severe.
5.0	Severe.
5.5	Moderate.
5.8	Slight.
6.0	Stable water **
6.6	No scaling, No tendency to dissolve scale.
7.0	No scaling. Moderate tendency to dissolve scale.
8.0	No scaling. Moderate tendency to dissolve scale.
9 0	No scaling. Strong tendency to descale.
10.0	No scaling, very strong tendency to dissolve scale.

TABLE 5.2

Factor A for Total Dissolved Solids

Total solids, ppm	Value of A
50	0.07
75	0.08
100	0.10
150	0.11
200	0.13
300	0.14
400	0.16
600	0.18
800	0.19
1000	0.20
1500	0.21
2000	0.22
2500	0.23
3000	0.24
4000	0.25
5000	0.26

TABLE 5.3

°F, tens	Factor B for Temperature °F, units: 0	2	4	6	8
30	---	2.6	2.57	2.54	2.51
40	2.48	2.45	2.43	2.40	2.37
50	2.34	2.31	2.28	2.25	2.22
60	2.20	2.17	2.14	2.11	2.09
70	2.06	2.04	2.03	2.00	1.97
80	1.95	1.92	1.90	1.88	1.86
90	1.84	1.82	1.80	1.78	1.76
100	1.74	1.72	1.71	1.69	1.67
110	1.65	1.64	1.62	1.60	1.58
120	1.57	1.55	1.53	1.51	1.50
130	1.48	1.46	1.44	1.43	1.41
140	1.40	1.38	1.37	1.35	1.34
150	1.32	1.31	1.29	1.28	1.27
160	1.26	1.24	1.23	1.22	1.21
170	1.19	1.18	1.17	1.16	---

TABLE 5.4

Factor C		for Calcium Hardness (as ppm CaCO$_3$)							

ppm					ppm units				
tens	0	1	2	3	4	5	6	7	8	9
0	---	---	---	0.08	0.02	0.30	0.38	0.45	0.51	0.56
10	0.60	0.64	0.68	0.72	0.75	0.78	0.81	0.83	0.68	0.88
20	0.90	0.92	0.94	0.95	0.96	1.01	1.02	1.03	1.05	1.06
30	1.06	1.09	1.11	1.12	1.13	1.14	1.15	1.16	1.17	1.19
40	1.20	1.21	1.23	1.25	1.25	1.25	1.26	1.27	1.28	1.29
50	1.30	1.31	1.32	1.33	1.34	1.34	1.35	1.36	1.37	1.37
60	1.38	1.39	1.39	1.40	1.41	1.42	1.42	1.43	1.43	1.44
70	1.45	1.45	1.46	1.47	1.47	1.48	1.48	1.49	1.49	1.50
80	1.51	1.51	1.52	1.53	1.53	1.53	1.54	1.54	1.55	1.55
90	1.56	1.56	1.57	1.57	1.58	1.58	1.58	1.59	1.59	1.60
100	1.60	1.61	1.61	1.61	1.62	1.62	1.63	1.63	1.64	1.65
110	1.54	1.65	1.65	1.65	1.65	1.66	1.67	1.67	1.67	1.68
120	1.68	1.68	1.69	1.70	1.70	1.70	1.70	1.71	1.71	1.71
130	1.72	1.72	1.72	1.73	1.73	1.73	1.74	1.74	1.74	1.75
140	1.75	1.75	1.75	1.76	1.76	1.77	1.77	1.77	1.77	1.78
150	1.78	1.78	1.78	1.80	1.80	1.80	1.80	1.80	1.80	1.80
160	1.81	1.81	1.81	1.81	1.82	1.82	1.82	1.82	1.83	1.83
170	1.83	1.84	1.84	1.84	1.84	1.85	1.85	1.85	1.85	1.85
180	1.86	1.86	1.86	1.86	1.87	1.87	1.87	1.87	1.88	1.88
190	1.88	1.88	1.89	1.89	1.89	1.89	1.89	1.90	1.90	1.90
200	1.90	1.91	1.91	1.91	1.91	1.91	1.91	1.92	1.92	1.92

ppm hundreds										
200	---	1.92	1.94	1.96	1.98	2.00	2.02	2.03	2.05	2.06
300	2.08	2.09	2.11	2.12	2.13	2.15	2.16	2.17	2.18	2.19
400	2.20	2.21	2.23	2.25	2.25	2.26	2.26	2.27	2.28	2.29
500	2.30	2.31	2.32	2.33	2.34	2.34	2.35	2.36	2.37	2.37
600	2.38	2.39	2.39	2.40	2.41	2.42	2.42	2.43	2.44	2.44
700	2.45	2.45	2.45	2.47	2.47	2.48	2.48	2.49	2.49	2.50
800	2.51	2.51	2.52	2.52	2.53	2.53	2.54	2.54	2.55	2.55
900	2.56	2.56	2.56	2.57	2.57	2.58	2.58	2.59	2.59	2.60

Use upper portion of table to find value of C for 3 to 209 ppm calcium hardness, lower portion for 210 to 990 ppm. Example: for 144 ppm calcium hardness (as CaCO$_3$) C = 1.76.

TABLE 5.5

Factor $\underline{D}$ for Alkalinity (as ppm CaCO$_3$)

ppm tens	0	1	2	3	4	5	6	7	8	9
0	---	0.00	0.30	0.48	0.60	0.70	0.78	0.85	0.90	0.95
10	1.00	1.04	1.08	1.11	1.15	1.18	1.20	1.23	1.26	1.29
20	1.30	1.32	1.34	1.36	1.38	1.40	1.42	1.43	1.45	1.46
30	1.48	1.49	1.51	1.52	1.53	1.54	1.56	1.57	1.58	1.59
40	1.60	1.61	1.62	1.63	1.64	1.65	1.67	1.67	1.68	1.69
50	1.70	1.71	1.72	1.72	1.73	1.74	1.75	1.75	1.76	1.77
60	1.78	1.79	1.80	1.81	1.81	1.82	1.83	1.83	1.83	1.84
70	1.85	1.85	1.86	1.86	1.87	1.88	1.88	1.89	1.89	1.90
80	1.90	1.91	1.91	1.92	1.92	1.93	1.93	1.94	1.94	1.95
90	1.95	1.96	1.96	1.97	1.97	1.98	1.98	1.99	1.99	2.00
100	2.00	2.00	2.01	2.01	2.02	2.02	2.03	2.03	2.03	2.04
110	2.04	2.05	2.05	2.05	2.05	2.06	2.06	2.07	2.07	2.08
120	2.08	2.08	2.09	2.09	2.09	2.10	2.10	2.10	2.11	2.11
130	2.11	2.12	2.12	2.12	2.13	2.13	2.13	2.14	4.14	3.14
140	2.14	2.15	2.15	2.16	2.16	2.16	2.16	2.17	2.17	2.17
150	2.18	2.18	2.18	2.18	2.19	2.19	2.19	2.20	2.20	2.20
160	2.20	2.21	2.21	2.21	2.21	2.22	2.22	2.23	2.23	2.23
170	2.23	2.23	2.23	2.21	2.21	2.24	2.24	2.25	2.25	2.26
180	2.26	2.26	2.26	2.26	2.26	2.27	2.27	2.27	2.27	2.28
190	2.28	2.28	2.28	2.29	2.29	2.29	2.29	2.29	2.30	2.30
200	2.30	2.30	2.30	2.31	2.31	2.31	2.31	2.32	2.32	2.32

ppm 100's	0	10	20	30	40	50	60	70	80	90
200	--	2.32	2.34	2.36	2.38	2.40	2.42	2.43	2.45	2.45
300	2.48	2.49	2.51	2.52	2.53	2.51	2.56	2.57	2.58	2.59
400	2.60	2.61	2.62	2.63	2.64	2.65	2.66	2.67	2.68	2.69
500	2.70	2.71	2.72	2.72	2.73	2.74	2.75	2.76	2.76	2.77
600	2.78	2.79	2.80	2.81	2.81	2.81	2.82	2.83	2.83	2.84
700	2.85	2.85	2.86	2.86	2.87	2.88.	2.88	2.89	2.89	2.90
800	2.90	2.91	2.91	2.92	2.92	2.03	2.93	2.94	2.94	2.95

TABLE 5.6

Factor E Equilibrium pH Value (pH$_{eq}$) Determined from Total Alkalinity*									
Alkalinity ppm	Alkalinity, ppm CaCO$_3$ tens								
hundreds 0	10	20	30	40	50	60	70	80	90
0 ---	6.00	6.45	6.70	6.89	7.03	7.14	7.24	7.33	7.40
100 7.47	7.53	7.59	7.64	7.68	7.73	7.77	7.81	7.84	7.88
200 7.91	7.94	7.97	8.00	8.03	8.05	8.08	8.10	8.15	8.15
300 8.17	8.19	8.21	8.23	8.25	8.27	8.29	8.30	8.32	8.34
400 8.35	8.37	8.38	8.40	8.41	8.43	8.44	8.46	8.47	8.48
500 8.49	8.51	8.52	8.53	8.54	8.56	8.57	8.58	8.59	8.60
600 8.61	8.62	8.63	8.64	8.65	9.66	8.67	8.67	8.68	8.70
700 8.71	8.72	8.73	8.74	8.74	8.75	8.76	8.77	8.78	9.79
800 8.79	8.80	8.81	8.82	8.82	8.83	8.84	8.85	8.85	8.86
900 8.87	8.88	8.88	8.89	8.90	8.90	8.91	8.92	8.92	8.93

Predictable Tower Scaling Profile

JOB NAME: _____ OPERATOR: _____ DATE: _____

TDS _____   Factor A _____   CALCIUM HARDNESS _____   Factor C _____
(Measured)   (Table 5.2)   (ppm hardness as CaCO$_3$)   (Table 5.4)
TEMP. ___°F   Factor B _____   TOTAL ALKALINITY _____   Factor D _____
   First Total _____   Second Total _____
   Plus __9.3__
   Third Total _____
Minus (Second Total) _____
   Fourth Total _____   x 2 = _____ pH Saturated
Minus TOTAL ALKALINITY (CaCO$_3$) Factor E _____
   Fifth Total _____   *See Table (5.1) to determine scaling severity.*

Scaling Severity is: _____

FIGURE 5.1 Predictable Tower Scaling Profile.

TOWER EQUILIBRIUM CONTROL

Environmental regulations regarding the discharge of cooling tower water are becoming more rigorous each day, and the cost of treating water, and discharging it into the sewer has also become more costly. Reducing water consumption is becoming more and more of an economical reality. Using the conventional method for treating cooling towers requires more water and more chemical whereas using the *tower equilibrium control* process, reduces the water usage in the tower and in the draining process.

Because of the high cost of water and chemicals along with increased environmental regulations some new methods of controlling scale and corrosion in cooling towers had to be developed. One of these developments is called *tower equilibrium control* or the *oxidation-reduction potential* (ORP) method of treating cooling towers.

Using the ORP method to control cooling tower waters was patented by HERC Products, Inc. The ORP method of water treatment can provide the user with an economical alternative to conventional treatment methods. The ORP method, which utilizes specific chemistry associated with ORP control can increase the cycles of concentration in cooling towers to high levels of TDS and in some cases it can achieve almost zero blowdown.

With use of the ORP method to treat cooling tower water, the cycles in the tower can be increased to very high levels of concentration. By reducing the blowdown, much less makeup water will be used. Using less makeup water, because of the reduced blowdown, requires less chemical as it is only necessary to treat the water being added to a system. It is to the benefit of the owner-operator to use less water and less chemical when treating the cooling tower water system. A short description of the process involved in this type of application is given in the following section.

OXIDATION-REDUCTION POTENTIAL

This is a measurement of the potential of electrons, in millivolts. The loss or gain of electrons results in an oxidation reduction reaction. Since most heat-transfer systems are constructed of metal, utilizing ever-changing water, there is typically an undesired equilibrium created. This measurement can allow control of electrochemical equilibrium.

In many chemical reactions electrons are transferred from one substance to another. By definition when a substance gives up electrons it is called an oxidation reaction, and conversely when electrons are added a reduction reaction has taken place. Actually oxidation and reduction reactions occur together at the same time. The available electrons from an oxidized substance are either given up instantly or are taken up by the reduced substance until an equilibrium condition has been reached.

The relative tendency of different substances to gain electrons varies with the number of electrons in the outer shell of the atom as well as the size of the atom or ion. Since it is impossible to measure absolute potentials, an arbitrary standard, the hydrogen ion, is chosen. Oxidation-reduction potentials are defined relative to this standard. Accordingly these substances can be tabulated in descending order with the substances that most easily gain electrons at the top.

Because oxidation and reduction reactions involve the transfer of electrons in the outer orbits of the atom, the relative tendencies for materials to oxidize or to reduce is dependent upon their atomic number, the number of electrons in the outer orbital, the pH of the reactive solution, and other factors. All of these tendencies can be measured in terms of the millivolt potentials measured by ORP instrumentation.

An ORP measurement is made using the millivolt mode of a pH meter. By substituting a metallic electrode for the pH glass electrode, many other ions beside the hydrogen ion can be detected with the same pH meter.

ORP measurements are used to monitor chemical reactions, quantify ion activity, and determine the oxidizing or reducing properties of the solutions being measured. While ORP measurements are somewhat similar to those of pH the potential value must be interpreted very carefully for meaningful results.

Because the ORP is characteristic of reactions involving both oxidation and reduction at the same time, it will vary as a function of the following:

1.   The standard potential $\underline{E}$ associated with each reaction
2.   Relative ion concentrations
3.   The number of electrons transferred in the reactions

Note that temperature compensation is not necessary.

In solutions that contain a host of constituents that oxidize and reduce simultaneously, *ORP relates to the concentration and activities of all the participating reactions.*

Not all chemical reactions involve ORP. For example, when sodium hydroxide reacts with the hydrogen chloride to form sodium chloride and water, there is no exchange of electrons between the atoms:

$$NaOH + HCl \rightarrow\rightarrow\rightarrow NaCl + H_2O$$

These reactions are called *ionic chemical reactions*.

Calcium ions combine with carbonate ions to form calcium carbonate. Calcium carbonate can concentrate beyond its solubility to form a sticky precipitate, which clings to pipe in the form of scale. There is no chemical reaction between the calcium ion and the carbonate ion in a precipitation condition. It is simply a precipitation due to the compound's limited solubility. As scale precipitates it would increase the ORP of the system water because of normally observed changes in the pH. However, as soluble calcium carbonate, kept in solution and concentrated, the ORP value would decrease.

EXAMPLE

As an example of the net oxidation- reduction reaction, when metallic iron reacts with oxygen to form iron oxide, as shown in the equation below, there is a transfer of electrons from the iron (Fe) atoms to the oxygen molecule to yield a ferric ion and an oxide ion. Since iron gives up three electrons and oxygen can take on only two electrons it is necessary to have four iron atoms react with six oxygen atoms (three oxygen molecules) in order for a total of twelve electrons to be transferred from the iron atoms to the oxygen atoms to form a ferric oxide and balance the equation.

$$4Fe + 3O_2 \rightarrow 2Fe_2O_3$$

A water treatment chemistry that contains a chemical species that will participate with the dissolved oxygen in an oxidation-reduction reaction must be used. This chemical is detected by the ORP electrode within the basic operating parameters of the cooling-water pH range of 7.0 to 9.0.

The responsive chemistry is the reaction product of a weak organic acid and a weak organic amine (base) used together. This is referred to as a soap compound or as a quaternary nitrogen (amine) compound.

As long as the soap or quaternary amine compound is present, the equilibrium will generate the organic acid to control the scaling components whether precipitated or in solution. These reactions allow for increased saturation limitations of alkalinity, calcium, magnesium hardness, or total hardness, and salts allow TDS levels to climb without untoward effects.

When the soap or quaternary nitrogen compound is depleted through consumption of the weak acid in the equilibrium to the point where the ORP reading is below the set point established for a threshold reading, additional chemical is added to the tower water. The chemistry is dosed in a concentration sufficient to react with any hardness from the makeup and to remove most preexisting scale formations or corrosion products, including silica as SiO_2 to greater than 300 ppm.

Although this is a new process, it shows great advantages over water treatment as we see it today.

CHAPTER 6

CONTROL DEVICES

One of the first cooling tower controllers could have been a slave overseer with a whip who would command a slave to move palm leaves, slower or faster, over the master's head. The moving palm leaves would then cause a breeze, increasing the evaporation rate on his master's body, cooling him down.

With the development of the cooling tower the first control for bleed was just to open a bleed valve and drain out some of the water at a predetermined rate. This bleed rate was estimated to be about 1 gallon per minute per ton.

As time went by it was discovered that with higher concentrations in the cooling tower water, more electricity was able to flow through the water. This was because as the concentrations increased, more salts were available, and with more salts conduction of electricity was easier. With less resistance in the water, because of the increased salts, more electrical current could flow through the water, and this is where *Ohm's* law then comes into action. *(See Chapter 3, section entitled "Conductivity.")*

It was not long before someone invented a control device to measure the conductance of the cooling tower water. When the cycles of water in the tower built up, the controller then measured the amount of electrical conductance in the water. When that conductance reached a predetermined set point a relay, located in the controller, would close, sending a voltage to a solenoid bleed valve at the tower. The bleed valve then opened, which allowed water to bleed from the tower to drain, lowering the level of water in the tower. The tower float control valve sensed this low level and opened, bringing fresh water into the tower basin. This fresh water caused the conductance in the tower water to increase, which in turn shut off the bleed valve, and the tower then filled up back to its original level.

If the bleed can be controlled, why not add a chemical feed pump to the controller that would then add chemical to the tower at the same time that tower was bleeding? By doing this the chemical would treat the incoming water at the same rate that the concentrated water was being bleed from the tower.

It was also found that any time that biocide was added to the tower water the conductance of the tower water increased very rapidly, opening the bleed valve. This caused the biocide to be drained out of the tower before it could do its job. Another control device was then incorporated. This new device would not allow the bleed valve to open during the time biocide was being added to the tower. This device was actually a seven-day timer that controlled not only the length of time the biocide remained in the tower, but the time and day that the biocide would be fed to the tower.

The tower controller was now becoming more complicated but it was getting better. Another timer was added to control the length of time that the biocide pump would operate. Then solid state electronics was added, "solid" state meaning use of printed circuits, etc. As time went on, pH control and then flow switches were added to control the on-off action of the controller. Temperature control for the total dissolved solids (TDS) probe, linear and digital readouts, tape recorders etc., were also added until the controller became not only very complex, but very efficient and easy to operate.

Microprocessors are controllers that are programmed like computers and can do just about anything that is asked of them. It just depends on how much money a company can afford to spend as to what the microprocessor will do.

DIGITAL READOUT VERSUS LINEAR READOUT

There are two ways of visually interpreting the readout of a controller. One is an electronic meter with a moving needle that points to the appropriate numbers. This is called a *linear readout* meter. The other method is an electronic device that indicates the appropriate number only, showing it as digits in a small window. This is called a *digital readout (See Fig. 6.1)*.

With linear readout the operator can see at a glance any change in movement that is taking place and how fast that change is occurring by viewing the swing of the needle or the rising or falling of the needle. A linear readout scale is harder to read due to the fact that the operator first has to find the range of the numbers and then read between the numbers and interpolate the number that the needle is at between those numbers.

With the digital readout the operator only reads the numbers that are actually shown on the display. New sets of numbers just keep appearing on the readout display. The operator cannot visually see how far or how fast the digital readout is changing. The digital readout is easier and quicker to see because the operator only looks at the display to read the exact number.

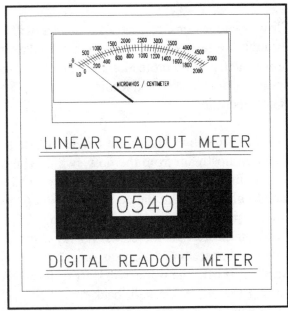

FIGURE 6.1
Linear and digital readout meters.

TEMPERATURE-COMPENSATED SENSORS

Temperature compensated TDS sensors give only slightly better accuracy for the dissolved solids measurement than non-temperature-compensated probes. Because *"temperature compensated"* sounds better and is used in all of the better new controllers, use it. It really does not make that much difference in actual practice.

FLOW SWITCHES

Flow switches are devices that signal when the water in a system is flowing. Some switches are more sensitive than others and the materials of construction in some are designed to withstand very low pH levels, while others will disintegrate at low pH levels.

Flow switches ensure that a controller will shut down when flow through the flow switch is below a set gallon/minute flow rate. The flow switch is good protection for a low initial investment. It prevents systems from bleeding and feeding chemicals when the cooling tower is not operating.

Caution: Caution should be taken so that a bleed valve is never installed upstream from the flow switch. If a bleed valve is installed upstream of the flow switch and the controller calls for a bleed, the bleed valve will take the complete flow of the water from the side-stream system. With no water flowing through the flow switch it would indicate that there was no flow and the system would shut down. Almost immediately, as soon as the power to the bleed valve was turned off, the water will start to flow through the flow switch, turning the control system back on. This will again turn the bleed valve back on and the sequence will repeat, constantly starting and stopping the controller. The above action could cause the bleed valve to open and shut rapidly or oscillate, making the controller ineffective.

CORPORATION STOP

A corporation stop is a device used for injecting a low-pH chemical into the main water piping system. It was designed for injecting acid or a low-pH chemical into the system so that

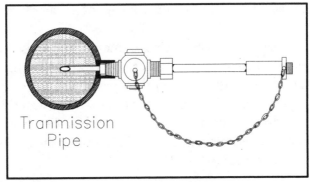

FIGURE 6.2   Corporation stop.

the low-pH would not destroy the piping. If chemicals with low-pH such as acid, are introduced into a piping system without using a corporation stop the low-pH chemical (acid) would enter the pipe and flow in a very concentrated solution along the inside walls of the pipe until the low-pH chemical was incorporated into the main flow of water. This constant exposure to a low-pH would cause the pipe to corrode and deteriorate at that location over which the low-pH chemical flowed.

A corporation stop screws into the side bung on the main pipe (See Fig. 6.2). A small stainless steel or PVC probe is forced through the corporation stop and into the center of the main water piping stream through a shutoff ball valve. When the low-pH liquid is injected through the corporation stop to the center of the pipe it is carried downstream, in the center of the water flow, mixing with the water as it goes. Because the low-pH liquid is mixing with the main water flow, it becomes less and less corrosive.

A corporation stop is designed with a ball-valve-type shutoff. When removing the probe from the main water line the corporation stop probe is only partially removed from the main pipeline. At that time the ball valve can be closed completely, preventing any loss of fluid from the main line. Then the probe is completely removed.

If a low-pH chemical is not being used, it is not necessary to use a corporation stop. Corporation stops were only designed to inject low-pH materials into the main stream of a water system.

The size of the opening in a corporation stop is very small, about 3/8 inches in diameter or smaller. Thus with any quantity of water flow, the pressure drop across the corporation stop is very high. A high-pressure drop in turn restricts the flow of water passing through the corporation stop so that the flow (in feet per second) through a controller is lowered to the point that there might not be enough flow to operate a flow switch.

Never terminate a sidestream of water that is used for control purposes with a corporation stop. A corporation stop is really designed to feed a low-pH chemical into a larger pipe with flowing water. It should not be used at the end of a bypass circulating line that is feeding back into the main circulating pipeline.

INJECTOR QUILL

An injection quill operates in the same way as a corporation stop, except that the quill cannot be removed from the main piping without draining the water from the main pipe.

COUPON HOLDERS

A coupon holder is a device for holding metal coupons in a stream of water. Metal coupons are used to measure the amount of corrosion taking place on the inside of a piping system. The coupons can also tell how well the system is being

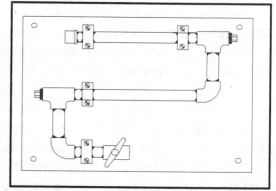

FIGURE 6.3   Corrosion coupon rack.

treated. Testing coupons is a powerful tool and can provide data that a knowledgeable and experienced water treatment engineer can use to make reliable predictions in the field. One advantage of coupons is that they can be given to a third party for evaluation.

Coupon holders for open systems, such as a condenser water system, are usually made of schedule 80 PVC pipe. If a coupon holder is to be installed in a hot-water system, schedule 80 PVC pipe fittings should not be considered. PVC schedule 80 pipe will begin to soften at temperatures above 140° F. Coupon holders made of steel should be used for installing coupons in hot-water systems (See Figs. 6.3 and 6.4).

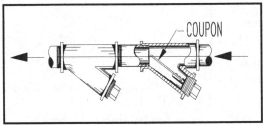

FIGURE 6.4

Closed system coupon holder.

Testing of coupons should be done every 60 days in order to make an evaluation of the water treatment results and any corrosion taking place in the system. An operator can have the coupons tested by a third party or they can be tested by your water treatment representative. In both cases, you should receive a complete analysis of the coupons.

WATER METERS

In large cooling tower installations, considerable cost savings can be achieved for the sewage bill if the water being fed to the cooling tower is measured and recorded with a water meter and the amount of bleed going to the sewer is measured and recorded with a water meter. If you subtract the gallons used for bleed from the amount of water used by the tower, the result is the amount of water evaporated. Since most of the water in the cooling tower evaporates the plant should not be assessed a sewer charge for that portion of water as it did not go down the sewer. The plant should only be charged for the amount of water going to the sewer.

POT FEEDER OR BYPASS FEEDER

Pot feeders (bypass feeders) are also a means of adding chemicals to closed systems. After a closed system has been treated and brought into the chemical treatment range, the system will require very little additives to maintain its chemical level. A two-gallon pot feeder, as a rule, is large enough for most buildings. The flow of water through the pot feeder should always enter at the bottom connection (IN) and exit at the top connection (OUT) (See Fig 6.5). If a

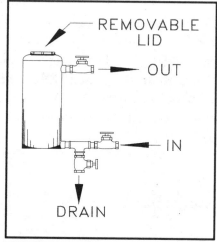

FIGURE 6.5
Pot feeder (bypass feeder).

powder-type chemical is being used, the force of the water entering at the bottom of the pot feeder will lift the chemical up and mix it with water, before ejecting it out of the top connection of the feeder.

When opening the lid on a pot feeder, always make sure that the pressure on the inside of the feeder is at atmospheric pressure and that the pot feeder is also isolated from the system being treated. This is done by opening a drain valve at the bottom of the feeder and releasing any pressure in the feeder to atmosphere. A pot feeder has a wide opening at the top for adding chemicals so a funnel is not required.

As a caution, it is always best to drain and isolate a pot feeder from the main system when the pot feeder is not being used to add chemical. By doing this, you save energy and the pot feeder is always empty when it is ready to be used again.

FILTER FEEDER

This is a device that serves two functions. One is to add chemical treatment to the inside of a closed system. The other is to remove solids from the inside of a closed system (See Fig 6.6).

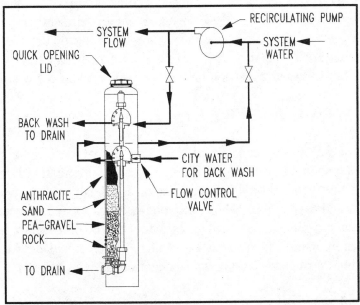

FIGURE 6.6   Filter feeder.

A percentage of the water in a closed system is always passed through the *filter feeder*. The filter cleans the water and then passes the cleaned water back into the closed system. Over a period of time all of the water in the system will have passed through the filter cleaning the system.

A filter feeder is primarily a small sand filter that can be backwashed, from time to time, to remove any dirt that has been collected from the closed system.

Chemical can be added to the closed system through a quick-opening lid, such as that found on a pot feeder. The lid is removed and then chemical is added as required, the lid is replaced, and the system is returned to service. This does not take the place of treatment. The closed system still must be treated with the proper chemicals.

CHEMICAL MIXING TANKS

Almost all commercial tower treatments come premixed in their own shipping containers so that these chemicals can be fed neatly right into the tower. If powdered chemicals are to be used, or if several chemicals are to be mixed, a mixing tank and an agitator will be required; otherwise a mixing tank and mixers are not really needed.

STEAM BOILER DEVICES

For process boilers, because of the rapid use of water, the boiler feedwater should be pretreated and softened as a minimum. For best results the water should be demineralized. Because all water has oxygen and carbon dioxide or other occluded gases in it, a means of removing oxygen and occluded gases should first be considered before the water is introduced into the boiler. Mechanical devices for liberating these gases are known as deaerators (See Chap. 9)

Although deaerators can remove most of these gases, some oxygen always gets into the boiler. As a last resort to remove the final traces of oxygen before the water is pumped into the boiler, sulfite is fed into the deaerator to remove any remaining oxygen before the water is fed into the boilers. Sulfite reacts with the oxygen to form a sulfate sludge, which can be removed from the deaerator by blowdown.

Sulfite should be fed into the boiler using a pulse-type water meter in the makeup water flowing to the deaerator. As the water passes through a water meter in the line to the deaerator the water meter sends a pulse to a chemical feed pump initiating a timed period for pumping sulfite into the deaerator. This system usually includes a chemical mixing tank with a chemical pump mounted on a stand, and the system is entirely separated from any other chemical feed system going to the boiler.

Each boiler in the system should have its own chemical feed system. If there are three boilers, there should be three separate chemical feed systems, one for each boiler. The best way to feed the chemicals is to feed them in direct ratio to the amount of feedwater being fed into the boilers. This is done by tying an auxiliary contact into the feedwater pump, so that as the boiler receives water it is also receiving chemical. The chemical pump should be adjustable so that the operator can control the amount of chemical being fed into the boiler.

BOILER ACCESSORIES

All devices that are directly connected with a boiler or that facilitate its operation and maintenance are called *boiler accessories*. The accessories usually provided are steam-pressure gages, water columns, feedwater regulators, safety valves, fusible plugs and blowoffs.

The water column and attached gage glass are one of the most important accessories on a steam boiler. When an operator walks into a boiler room, he or she can tell if the boiler is a steam boiler or a hot-water boiler by noticing if the boiler has a water column and gage glass. If the boiler has these accessories, he or she will know that the boiler is a steam boiler. It is also the first thing an operator looks at to see if there is water in the boiler.

The water column is connected to the front of the boiler drum, so that the water level may be visible from the boiler-room floor. The top of the water column is connected with the steam space and the bottom of the water column to the water space in the boiler drum.

In many cases, a float valve is incorporated into the water column, and the float operates a mercury switch that is activated when the water level is low. This switch in turn operates the feedwater pump, which pumps feedwater into the boiler, maintaining the proper water level in the boiler.

FUSIBLE PLUGS

Boilers (especially fire-tube types) operating at less than 225 lb/in^2 pressure are generally protected by fusible plugs. Fusible plugs consist of steel or bronze bushings, filled with a tin alloy that melts at or about 450°F. They are inserted into the boiler drum at the lowest permissible water level location as listed by the American Society of Mechanical Engineers (ASME) boiler code. The melting point of tin is above the temperature of the steam and below the temperature of the hot gases. The small end of the plug is exposed to the products of combustion. When the water level is low enough to uncover the plug, the alloy melts and steam escapes. In this way, probable excessive boiler pressure is prevented and the escaping steam attracts the attention of the operator, so that precautions may be taken to prevent overheating of the boiler metal.

SAFETY VALVES

Safety valves are used to protect boilers against excessive steam pressure by opening automatically at a desired pressure and allowing the steam to escape. The number and size of safety valves for a given boiler are usually specified by city or state legislation or by insurance companies (See Chap. 9, section entitled "Safety Valves".)

CHAPTER 7

HYDRONIC SYSTEMS

Water systems that convey heat to, or from, a conditioned space or process with hot or chilled water are frequently called *hydronic systems*. The water in these systems flows through piping that connects to a boiler, water heater, or chiller, or to a suitable terminal heat-transfer unit, located in the space or process.

A closed water system is defined as a system with no more than one point of interface with a compressible gas or surface, and the entire piping system is always filled with water or a heat-transfer fluid. These closed systems are classified by operating temperatures and are discussed in the following.

LOW-TEMPERATURE WATER SYSTEMS

These systems operate within the pressure and temperature limits of the ASME Boiler Code for low-pressure boilers. The maximum allowable working pressure for low-pressure hot-water-heating boilers is 160 lb/in² gage with a maximum temperature limitation of 2000F. The usual maximum working pressure for a low-temperature water (LTW) system is 15 lb/in² gage. *Steam-to-Water* or *Water-to-Water* heat exchangers are also used for heating low-temperature water.

MEDIUM-TEMPERATURE WATER SYSTEMS

These systems operate at temperatures between 250°F and 350°F, with a pressure rating for boilers and equipment of 150 lb/in² gage.

HIGH-TEMPERATURE WATER SYSTEMS

High-temperature water systems operate at temperatures above 250°F and at operating pressures above 160 lb/in² gage.

CHILLED-WATER SYSTEMS

Hydronic cooling systems normally operate with a design supply water temperature of 40 to 55°F and within a pressure range of 160 lb/in² gage. Antifreeze or brine solutions may be used for certain applications (usually process applications) that require temperatures below 40°F or for coil freeze protection.

CHILLED-WATER BRINE SYSTEM

This type of a system cools down a brine solution, usually in a holding tank, overnight to below 32°F. During the day the system circulates the cold brine solution throughout the building for cooling.

DUAL-TEMPERATURE WATER SYSTEMS

Dual-temperature water systems are a combination of heating and cooling systems. A dual system circulates both hot and/or chilled water through common piping and terminal-heat transfer apparatus. These systems operate within the pressure and temperature limits of low-temperature water systems, with usual winter design supply water temperatures of about 100 to 150°F and summer supply water temperatures of 40 to 45°F.

Low-temperature water systems are used in buildings ranging in size from small, single dwellings to very large and complex structures. Terminal-heat transfer units include convectors, cast iron radiators, baseboard and commercial finned-tube units, fan-coil units, unit heaters, unit ventilators, central station air-handling units, radiant panels, and snow-melting panels. A large storage tank may be included in the system for storing energy to use when heat input devices, such as the boiler or solar energy collector, are not supplying energy.

The one thing in common with all of the above noted systems is that they are called *closed systems*. Even though they are called closed systems, however, they all leak water. Sometimes hot water leaks through piping joints and flashes into steam, sometimes water is lost to the atmosphere through pump seals, and worst of all hot water is lost when an operator blows down a closed system.

NOTE:

A CLOSED SYSTEM SHOULD NEVER BE BLOWN DOWN.

When a closed system is blown down it must take on fresh water to maintain the system pressure. This fresh water being added to the system contains dissolved oxygen. Dissolved oxygen promotes corrosion, which is harmful to the inside of the piping.

Because of the water loss in a closed system there must be some way of adding fresh makeup water back into the system. One way is to have a pressure-reducing valve in the makeup water line to the system. Thus if the water pressure drops in the closed system, due to a loss of water, the pressure-reducing valve will sense this lower pressure, open a makeup valve, and add water back into the system until its operating pressure has been reached.

Another way to add water to a closed system is to use a pumping system that will add water when the water pressure in the closed system drops. This system is often referred to as a *glycol feed system*. Glycol, which is an antifreeze, is often mixed with the closed system water to prevent freezing of the water in the piping. Freezing water in pipes can cause the pipes to break. Some glycols are poisonous when consumed. There is always the danger of water mixed with glycol backing up into the domestic water, supply so the glycol feed system is often used to separate the domestic water completely from the closed system water.

GLYCOL FEED SYSTEMS

To get around the problem of a closed system, using glycol, mixing with or having a crossover between the closed system and the domestic water, a *glycol feed system* was developed. The glycol feed system is a device consisting of a plastic or fiberglass holding tank, a system-sensitive differential pressure switch, and a positive displacement pump for pumping a glycol mixture into the closed system. The glycol feed system is completely separated from the domestic water system by an air gap (See Fig. 7.1).

When the pressure in the hydronic system drops it triggers a pressure switch causing the positive displacement pump to operate. The pump then sends a mixture of glycol and water into the hydronic system, bringing the pressure in the system back to its operating parameters.

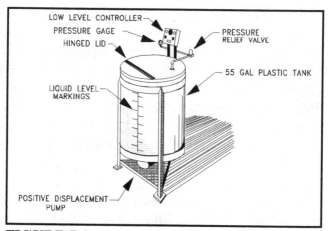

FIGURE 7.1   Glycol feed tank.

EXPANSION TANK

The one point of interface in a closed water system is the expansion chamber or so-called expansion tank. Half an expansion tank contains system water, with the remaining area in the tank consisting of air. In effect the system water is riding on a cushion of air. The pressure in an expansion tank is determined by the operating conditions of the system. This means that the air in the compression tank is compressed, maintaining a predetermined pressure in the closed system. The reason for this is that the net positive suction head (NPSH) of the system has to be more than the NPSH of the circulating pump in order for the pump to operate properly (see Chap.10, for an explanation of NPSH). This also prevents cavitation of the circulating pump and cushions any water hammer that might be present in the system.

The expansion tank sometimes becomes what is called *water logged.* This means that the air has leaked out of the expansion tank. Because now there is no air in the tank, compression of the air cannot take place. When this occurs the tank should be drained of water. This will again allow an air cushion to be placed on the top of the water in the expansion tank.

GLYCOL

As mentioned above some closed systems use antifreeze. There are two kinds of antifreeze used in closed systems: propylene and ethylene.

Propylene glycol is approved by the Food and Drug Administration (FDA) for use around food products and is not considered hazardous under ordinary conditions.

Ethylene glycol, on the other hand, can cause gastrointestinal irritation, nausea, vomiting, and diarrhea and is poisonous to the human body.

Some states have passed laws such that ethylene glycol cannot be used in hot- or chilled-water systems. This is because there is always the possibility that a crossover leak could occur between the domestic water system, supplying water to the hydronic systems, and the heating or chilled-water systems. If hot water or chilled water were to back up into the domestic water, it could poison the water.

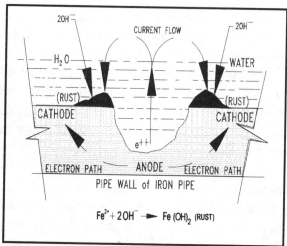

FIGURE 7.2   Corrosion.

CORROSION

Fouling on the inside of pipes, in closed water systems, is the greatest cause for hydronic equipment failure. It is this fouling in the form of iron or copper oxides that plugs strainers and control valves and reduces the heat-transfer surfaces of boilers and condensers. It is necessary to understand the corrosion process in order to solve fouling problems in a closed piping system.

Aqueous corrosion is an electrochemical process by which metals return to their original state. The electrochemical process can take place anywhere in a piping system. It is not necessary that there be two dissimilar metals in the system. Many metals, particularly iron, can have anodic and cathodic areas in the same piece of pipe and at the same time. In order for corrosion to take place, in a closed system, a corrosion cell must first exist. A corrosion cell consists of an anode, a cathode and an electrolyte. The metal ions in the steel pipe form the anode as shown in Fig. 7.2. When an iron pipe is immersed into an electrolyte, current will flow from the anodic to the cathodic areas and corrosion will result. The ions begin to dissolve into the electrolyte (*the water in the pipe*). As these ions dissolve they leave behind electrically charged particles called *electrons*.

Electrons always flow in the direction opposite to the electrical current, so when a neutral iron atom leaves the iron pipe, it loses two electrons and becomes an iron ion, Fe^{2+}. To complete the corrosion cell, there must be an electron path, and this path would be the steel pipe carrying the water. The electrons then work their way through the metal pipe and attach themselves to the cathode, which is at the surface on the inside of the pipe. At the cathode, these two Fe^{2+} electrons combine with two positively charged hydrogen irons and form a molecule of hydrogen gas. If this were all that happened, the system would reach equilibrium and corrosion would stop. In fact, this process is called *passivation* and is used to protect some metals. However, other reactions also take place in the electrolyte. The removal of the hydrogen ions from the electrolyte, at the cathode, leaves an excess of hydroxyl ions. These ions react with the iron, forming *ferrous hydroxide*. The ferrous hydroxide then precipitates, allowing the corrosion reaction to continue. The result is a loss of metal in the first action and a deposit in the second. This reaction between different sections of the same piece of metal or between two dissimilar metals in electrical contact with each other causes *corrosion*.

$$Fe^+ + 2OH^- \Rightarrow\Rightarrow\Rightarrow Fe\ (OH)_2$$

A simple explanation of the corrosion process is that *When an electrochemical reaction sets up in a closed piping system with a current flowing from the anode to the cathode, through the electrolyte, the iron in the pipe goes into solution, or corrodes.*

Pitting is frequently caused by the presence of chloride ions and is accelerated by the presence of oxygen in the electrolyte. Actually corrosion includes such things as the rusting of iron, rotting of wood, weathering of stone, etc. Aqueous corrosion of metal is always associated with an electrochemical reaction.

Galvanic corrosion is an accelerated corrosion of metal due to the coupling of two different galvanic potential metals that come together, such as a steel and copper flange. In a very intimate contact such as this, between two dissimilar metals, corrosion may accelerate between the two metals at a rapid pace. The more anodic material will corrode faster when coupled with a less anodic material (See Table 7.1). Actually all metals corrode, but certain metals corrode much faster when coupled or connected together with other metals. The more active the metal, the more rapidly it will corrode when coupled to a less active metal. The question is, "How much faster?"

Table 7.1

Electrode Potentials		
Element E (V)		
Ca	Calcium	2.87
Mg	Magnesium	2.37
Al	Aluminum	1.66
Mn	Manganese	1.18
Zn	Zinc	0.0763
Ch	Chromium	0.074
Fe	Iron	0.440
Cd	Cadmium	0.403
Co	Cobalt	0.277
Mo	Molybdenum	0.250
Sn	Tin	0.136
Pb	Lead	0.126
H	Hydrogen	0.0
Cu	Copper	-0.337
Ag	Silver	-0.800
Pl	Platinum	-1.2
Au	Gold	-1.5

Speed of Corrosion. How fast corrosion proceeds is related to the following:

1. The composition of the two coupled metals
2. The surface finish of the two coupled metals
3. The liquid flow rate past the two coupled metals
4. The dissolved oxidizing species in the solution (oxygen, ferric ions)
5. The species that tend to destroy the protective films (like chloride, sulfide, bromide)
6. The species that tend to form or to improve protective films (carbonate, phosphate, nitrite)
7. The relative areas of the two galvanic materials

Table 7.2 illustrates that the relative of corrosion between the coupled and uncoupled active material may vary considerably but there is little question that they will corrode more rapidly and fail prematurely when the two dissimilar metals are coupled together.

When oxygen is present the oxygen reacts with the ferrous hydroxide at the anode to produce ferric hydroxide, which is even more insoluble. The explanation above can also be applicable to steam boiler piping. To protect the internal piping in a steam boiler from corrosion, it is best to remove all oxygen before it enters the boiler.

TABLE 7.2 Rate of Corrosion.

Anode*	Cathode[+]
Aluminum	Iron
Aluminum	Copper and its alloys
Aluminum	Stainless Steel
Iron	Copper and its alloys
Iron	Stainless steel
Zinc	Iron
Zinc	Copper

* Corrodes more rapidly when connected.
+ Does not corrode when connected.

PROTECTION

Since the electrochemical cell is necessary for corrosion to take place, the logical method of preventing corrosion is to *destroy the cell*. If the anode can be isolated from the cathode by a dielectric or the anode and cathode can be isolated from the electrolyte with an *insulated film*, the corrosion process would stop.

The most fool proof method for breaking up a cell in which different metals are used would be to use a dielectric between the anode and cathode so as to prevent any current from passing between them (See Fig. 7.3).

Another method, one best suited for systems where two different metals are not used, would be to add a chemical to the inside of the closed piping system that will coat the pipe with an insulating film.

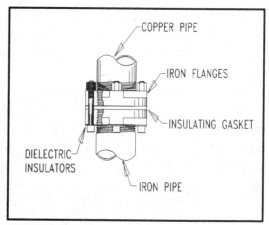

FIGURE 7.3 Dielectric flange.

The film should be thin enough that it will not affect heat transfer but will still insulate the cell, preventing it from passing electricity through the electrolyte path. This means that the water chemistry always has to be maintained with an insulating film on the inside of the pipe. The best solution for corrosion protection is to use both of the methods noted above.

Chemicals that control corrosion are called *corrosion inhibitors.* They are referred to as cathodic or anodic.

Cathodic	Anodic
Polyphosphate	Orthophosphate
Zinc	Chromate
Molybdate	Nitrite
Polysilicate	Ferrocyanide

Some corrosion inhibitors work better than others, and some have problems that overcome their desirability to be used.

Nitrite is an anodic inhibitor and in conjunction with other chemicals has proven to be an excellent inhibitor for closed systems. For nitrite to be effective it requires a high pH (above 7.5) and the concentration of nitrite in the fluid must be kept over 500 ppm, preferably in the 700 to 1000 ppm range.

Molybdates, a cathodic inhibitor, when used with other chemicals, has also proven to be a very good inhibitor for closed systems.

PITTING

Pitting is another form of corrosion. It is caused by the same electrochemical cell process as corrosion, only the corrosion process has been confined to a very small area of the pipe. Pitting is frequently caused by the presence of chloride ions and is accelerated by the presence of oxygen in the solution (See Fig. 7.2).

Example

If a closed water system is being treated at an inhibitor level of 500 ppm using a nitrite inhibitor and the inside of the pipe is corroding at the rate of 10 mils per year, it must be assumed that inhibitor level on the inside of the pipe was allowed to drop below 500 ppm.

Because of this the inhibitor is now only protecting 90% of the wall on the inside of the piping system. This means that iron is being removed from 10 % of the anode.

If corrosion is now being concentrated on this reduced wall area of the pipe and it is corroding at the rate of 100 mils per year in this small area, a severe pitting condition will arise.

It has also been found that pitting is more severe on the downstream, cool side of a heatexchanger or piping system. The corrosion rate doubles for every $10°F$ loss in temperature. This would also be true for chilled-water systems.

MEASUREMENT OF CORROSION

The corrosion rate of the metal in a water system can be measured by inserting a sample specimen, using the same material as the piping, into the flow of the water in the piping system. The most important step in corrosion testing is to completely identify the material to be tested. Chemical composition, fabrication data, metallurgical history, and positive specimen identification are required for meaningful results. Usually, two specimens, copper and iron, are sufficient.

Since the degree of scaling varies greatly on equipment and because the determination of the corrosion resistance of the metal itself is of prime importance, a clean metal surface is most often preferred. Rectangular specimens are usually preferred because they are easier to handle. The thickness of the coupons varies, but as a rule the coupon is sized at 1/16 in thick by ½ in wide by 2 in long.

Each specimen is cleaned and weighed, and this information is then recorded. The coupon is installed in the water system and left for a predetermined time, after which it is removed. The specimen should then be carefully examined for type and uniformity of surface attack such as etching, pitting, tarnish, film, and scale etc. If pitting is observed, the number, size, and distribution, as well as the general shape and uniformity, of the pits should be noted. The maximum and minimum depth of the pits can be measured with a calibrated microscope or by the use of a depth gage. Photographs of the specimens will serve as an excellent record of the surface appearance before and after.

After cleaning, the weight of each specimen should be determined to the nearest 0.1 mg on an analytical balance scale and the loss in weight calculated. The corrosion rate in mils per year can be calculated using the following equation:

Corrosion rate loss (mils per year) =

$$\frac{weight\ loss\ (g)\ \cdot\ 534.000\ \cdot\ hours\ of\ exposure\ (h)}{metal\ density\ (g/cm^3\ \cdot\ metal\ area\ (inch^2)}$$

For best results metal specimens should be weighed and looked at by a trained professional in the field of corrosion engineering.

FREE COOLING

Free cooling is a term given to a cooling system where the condenser water from the tower bypasses the chiller, going directly to the closed system. Because the cooling tower's cold water temperature drops as the load and ambient temperature drops, the water temperature eventually becomes low enough to serve the cooling load directly, allowing the energy-intensive chiller to be shut off and be bypassed. Free cooling allows the chiller to be bypassed with the tower water going directly to the cooling load.

Although free cooling can be an energy cost savings to the owner the primary disadvantage is that it allows the relatively dirty tower water to contaminate the clean chilled-water system. Because of this the owner could incur other costs that would be much higher than the energy cost savings of bypassing the condenser.

Cooling towers, because of their environment, act as air washers, loading the tower water with oxygen, other gases, and dirt that are washed into the tower from the incoming air.

Condenser water treatment is entirely different than closed system water treatment. When tower water is added to a closed system all kinds of chemical treatment changes in the water must be taken into account.

1.   Cooling tower water is mainly being treated to prevent scale from forming in the tower and in its associated piping. Dissolved solids in the tower water, when properly treated, can be allowed to build up to high levels before scale will form. Inhibitors used to prevent corrosion in cooling towers are different than the inhibitors used to treat corrosion in a closed system.

2.   When *free cooling* is being used, solids, oxygen and other gases are collected from the washed air in the tower water and are then dumped into the closed system. The majority of these

solids will have a specific gravity above 1 or higher. Some of the particle sizes will be much larger than orifices found in closed system control valves, which were never designed to cope with large particles which often cause the control valves to fail much of the time.

In high-rise buildings where the cooling tower is located on the top of the building and where *free cooling* is used the chilled-water header pipes drop down for many floors. The velocity of the condenser water through these pipes, in most cases, is not high enough to return or lift the solids, washed in through the tower, down through the headers, through the cooling load, and then back up to the top of the tower on the roof of the building. Because of this the particles of dirt and solids remain at the bottom of the headers, eventually accumulating with the solids to the point that they restrict the flow of water or/and in some cases they completely shut off the flow of water in the closed system. In addition, if this water is being processed through small computer circuits for cooling purposes, the particles will plug these circuits, causing computers to drop out of service because of high heat load.

3.   Due to the fact that a cooling tower is an air washer the incoming air also brings bacteria to the tower system. These bacteria then, because of its dark warm environment, form growths in the water, which also impede the flow of water back up to the tower.

Filtration systems, either sidestream or full flow, minimize this contamination to some extent, but many engineers consider this to be a concern. Free cooling systems that allow tower water to be pumped into and through a closed system are not recommended.

Other types of free cooling systems, such as the fluid cooler type of tower (Fig. 2.4) keep the tower water isolated from the closed system water. Tower water in this type of system can easily be chemically treated the same way as any other tower water system. The closed system water in the fluid cooler can also be chemically treated in the same way that any other closed system water is treated.

In addition to protecting a closed system from corrosion, precautions must be taken to protect the inside fluid of a closed circuit fluid cooler from freezing. This can be done by using an antifreeze solution.

CHAPTER 8

STEAM FACTS

Heating of water at any given pressure eventually causes the water to boil and release steam from the water. The heat required to bring water from 32° F (*base point for all water and steam properties*) to its boiling point is called the *enthalpy* of the liquid. Enthalpy is measured in Btu/lb.

When water boils it has the same temperature as steam. *Both water and steam have the same temperature at this point.* This point is called the *saturation temperature*. For each boiling pressure, there is only one saturation temperature. In other words for different pressures, there will be different saturation temperatures. When water reaches the saturation point the temperature remains constant even though heat (Btu) is still being added to the boiling water. The extra heat or Btu being added to the water is being used to change the water from the liquid state of the water to the vapor state, steam. As described above this extra heat is called the *enthalpy* (See Fig. 8.1).

The higher the pressure in a boiler system the more Btu it takes to bring the water to the boiling point. At the same time it takes less Btu to change the water from a liquid to the vapor state or steam.

To increase the temperature of the steam (*superheat*) more Btu in the form of heat must be added to the steam. Enthalpy of the steam always increases by the amount of Btu added per pound of steam.

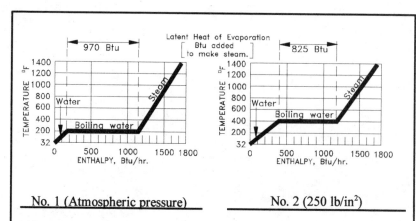

No. 1 (Atmospheric pressure)

When heat is applied to water at atmospheric pressure it will bring the water from 32°F to 21?F, the boiling point of water. Additional Btu of heat will then have to be continually added to the water until the enthalpy Btu reach about 1160 Btu before the boiling water then turns to steam. This means it took an additional 970 Btu to change the boiling water to steam. This is called Latent heat of evaporation.

No. 2 (250 lb/in²)

When heat is applied to water that is under 250 lb/in² of pressure, it brings the temperature from 32 to 400°F. It takes more Btu to convert the boiling water into steam. The latent heat of evaporation is not 825 Btu. A total of 1200 Btu was used to convert the water into steam. It took far fewer Btu to convert boiling water into steam when the water was under 250 lb/in² pressure.

FIGURE 8.1   The enthalpy process.

Since the properties of steam and water are definitively fixed by nature they can be measured and tabulated. The data most widely used in the United States are found in Thermodynamic Properties of Steam by Joseph H. Keenan and Frederic G. Keys. The definition for the production of steam is *the heating of water in a vessel until the water evaporates, producing steam.* When steam is withdrawn from a pressure vessel it is replaced with boiler feed water.

If a boiler is using all of the steam that it produces the boiler is called a *process boiler.* When a boiler receives most of its steam back in the form of condensate it is called a *heating boiler.* A process boiler requires a better quality of pretreated water than a steam heating boiler because a steam heating boiler receives back 80 to 90% of its processed steam, which is pure condensate water.

Boiler feedwater is different from *boiler makeup water.* Boiler feedwater is made up of both the returning condensate water and the makeup water as in a heating boiler. *For a process boiler, boiler makeup water is the boiler feedwater.* It is very important that this difference be understood as it has a great deal of bearing on the amount of chemical required, the amount of blowdown required, and the amount of makeup water required to sustain the boiler in operation.

If the makeup water added to the boiler is not pretreated but comes from a well or a municipality, it can contain large amounts of oxygen, calcium or silicon constitutes, inorganic salts, dissolved organics, and other gases. The amount of these solids in the boiler water affects the quality of the steam that is produced. Each solid particle that is in suspension in the boiler water will act as the nucleus for a drop of moisture when the boiler is operated at higher levels of steam production. The higher the concentration of these solids in the boiler water the "wetter the steam." Wet steam in a boiler is an undesirable product.

These impurities when heated in the boiler cause *hard scale* to form in the boiler. This *hard scale* is caused by calcium sulfate, calcium silicate, magnesium silicate, and silica. *Soft scale*, on the other hand, is derived from calcium bicarbonate, calcium carbonate, calcium hydroxide, magnesium bicarbonate, iron carbonate, and iron oxide. *Corrosion* in the boiler is caused by oxygen, carbon dioxide, magnesium chloride, hydrogen sulfite, magnesium sulfate, calcium chloride, magnesium nitrate, calcium nitrate, sodium chloride, and certain oils and other organic matter.

Makeup water added to the boiler can be externally treated before it ever gets into the boiler or it can be treated after it enters the boiler. After boiler feedwater enters the boiler it becomes very difficult to treat. For that reason the makeup water should be externally pretreated before it is ever introduced into the boiler. Pretreatment must be carefully selected and controlled to produce a high-quality feedwater.

External pretreatment could consist of filtration, water softening, demineralizing, reverse osmosis, etc. The incoming makeup water should always be analyzed by a water treatment company to determine what pretreatment of the makeup water should be done in order to make it acceptable for a particular type of boiler installation.

BLOWDOWN

To prevent the amount of solids in the steam boiler from becoming too high the boiler operator must blow down the boiler from time to time so as to remove these solids. Accurate records on the blowdown of the boiler should always be kept, so that the operator knows how much to increase or decrease the boiler blowdown (See Fig. 8.2).

Some boilers have blowdown valve ports located in the bottom and at the back of the boiler. Some have blowdown valve ports located at the bottom and at the front of the boiler, while other boilers have blowdown ports located on the bottom in the front, middle, and back of the boiler.

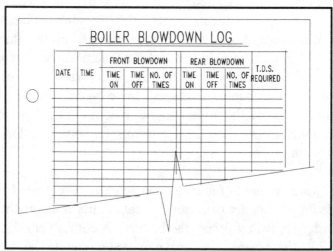

FIGURE 8.2 Typical steam boiler blowdown log.

BLOWING DOWN THE BOILER

Using the analogy of a bathtub, with the bathtub representing the bottom half of the steam boiler, imagine the following. With the bathtub half full of dirty bath water the bottom drain in the tub is opened. The water in the tub slowly drains out through the bottom drain, leaving a ring of dirty bath water and solids on the inside of the tub. If the water in the tub were to be agitated as it was draining it would pick up some of the solids on the side of the tub, taking them to the drain. There might also be a small circular whirlpool where the water leaves the tub and enters the drain.

If another bathtub were to be placed upside down over the first tub and then the two tubs were sealed together, they would become one vessel. By adding 15 lb of pressure to the inside of the two bathtubs they have become a pressure vessel.

When the bottom drain in the tub is opened under 15 lb/in^2 of pressure, what happens? The dirty bath water still leaves a ring around the tub as it is draining out. The one exception is that the water is now discharging under 15 lb/in^2 of pressure. The bath water is leaving the tub very rapidly and under a lot of force.

If the bathtub drain were to be stopped suddenly while the tub was draining at this rapid rate, the bath water would surge back into the tub, scrubbing the walls of the tub. This scrubbing action would also cause any solids in the bottom of the tub to go back into suspension. If the drain in the bottom of the bathtub were then opened again, the bath water would carry those suspended solids as well as any other solids that might have precipitated out and be flushed to the drain.

The same analogy can be used when blowing down a steam boiler. The operator should first open the boiler blowdown main shut-off valve going from the boiler to the quick-acting blowdown valve that is piped to the blowdown tank so that the water has a clear path to the blowdown tank.

The operator would then open the quick- acting knife-type blowdown valve, located between the boiler shut off valve and the boiler blowdown tank. With the knife valve open the operator would count 5 seconds and then rapidly close the knife valve, wait for another 5 seconds, and then repeat the opening and closing of the knife valve. The number of times that the valve is opened and closed should be recorded along with the time that the knife valve was left in the open position. If there are two blowdown ports, one in the front of the boiler and one in the back of the boiler, records should be kept of the number of blowdowns and the length of the blowdowns for each port.

SKIMMER BLOWDOWN

Blowdown of the dissolved solids in the boiler water can actually be controlled automatically by using a skimmer blowdown system (See Fig. 8.3). The skimmer controller periodically samples boiler water by opening the blowdown line for a short sampling period (See Fig. 8.4). If the total dissolved solids (TDS) is below the trip set point the blowdown valve closes after the timed sample has been taken. If the TDS exceed the preset trip set point, the controller overrides the timer, allowing the blowdown valve to remain open until the system solids return to the desired TDS set point. A conductivity meter that measures the TDS is used to control the amount of solids.

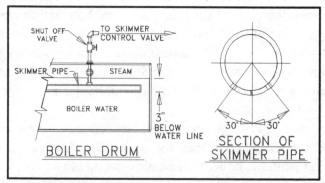

FIGURE 8.3   Typical skimmer blowdown pipe.

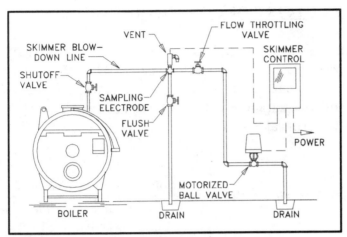

FIGURE 8.4 Typical skimmer control.

A blowdown skimmer should take water from the length of the boiler and about 3 in below the operating water level of the boiler. Water from the inside of the boiler is then collected in the skimmer line and fed to a skimmer blowdown valve on the outside of the boiler. In the event that the skimmer blowdown valve should ever fail in an open position, water would still remain in the boiler keeping the boiler tubes covered with water, protecting them from damage. In addition, the skimmer valve is a much smaller valve than the bottom blowdown valve so it takes a longer time for the water to escape from the boiler.

An automatic bottom blowdown valve should *never* be used for this type of service. If there were to be a failure of the automatic bottom blowdown valve in the open position, all of the water in the boiler would blow out, damaging the boiler tubes.

A bottom blowdown valve will always be necessary to remove the sludge, which has been precipitated out of the boiler water by the use of chemicals.

BOILER CYCLES OF CONCENTRATIONS

It is not always practical to measure the makeup water or the condensate water so as to determine the blowdown rate, but by measuring the chlorides (See chap. 3, section entitled "Chlorides"), cycles of concentrations in the boiler can be calculated by using the following formula:

$$Cycles \ of \ concentration \ = \ \frac{Boiler \ blowdown \ chlorides}{Feedwater \ chlorides}$$

A steam boiler should be operated at the highest possible number of cycles for economical reasons. Recommendations for this purpose have been made by the *American Boiler Manufacturers*. The following table shows the maximum TDS to be maintained in a boiler at different operating pressures.

TABLE 8.1

American Boiler Manufacturer's Standards

Operating pressure lb/in²	MAXIMUM TDS ppm
0 - 300	3500
301 - 450	3000
451 - 600	2500
601 - 750	2000
751 - 900	1500
901 - 1000	1250
1001 - 1500	1000

BOILER WATER ANALOGY

It is important for a plant operator to know the total amount of steam produced by the boiler, the amount of condensate being returned to the feedwater tank, the amount of makeup water required, the amount of boiler feedwater required for the boiler, and the amount of   steam loss throughout the system, as well as the amount of blowdown water used, to maintain the solids in the boiler at the proper level.

The plant operator should make a copy of Fig.8.5 and insert the characteristics of his or her boiler. It would prove to be a helpful tool should there be trouble with the boiler at a later date.

There are a lot of parts to the following   problems.   To avoid confusion, the first thing to do is get the facts organized in such a way that they are simple to understand. If we give an alphabetical letter to each of the quantities shown below it will be much easer to solve the problem.

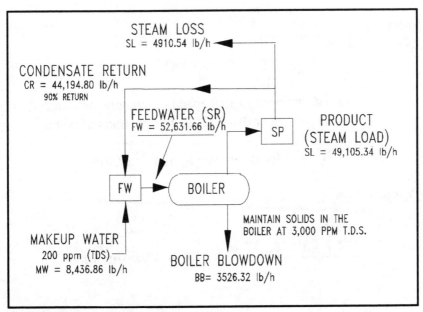

FIGURE 8.5   Boiler water analogy.

CALCULATIONS SHOWING HOW MUCH WATER IS IN A BOILER

Boiler Horsepower

Boiler horsepower can be calculated in several different ways.

1 boiler hp	=	10 ft² of boiler heating surface
1 boiler hp	=	34.5 lb of evaporated water/hr
1 boiler hp	=	4 gal of water evaporated/hr

Problem 1

If a boiler's water drum capacity holds 193.7 gal of water, how many pounds of water would that be?

Note: Pounds of steam = Pounds of water

Solution

$$BC \cdot 8.337 \text{ lb} = DC$$

where   BC = boiler drum capacity in gallons of water
DC = boiler drum capacity in pounds of water

(8.337 lb) is the weight of 1 ft² of water

Computation

$$\frac{193.7 \ gal \cdot 8.337 \ lb}{1 \ gal} = 1{,}614.87 \ lb \ of \ water$$

Remember to always convert units, such as *gal,* to the same type of unit, gal. Do not mix pounds of water with gallons of water.

Problem 2

If a boiler has a steaming rate of 30,000 lb of steam per hour and the feedwater contains 60 ppm undissolved solids, what would be the weight of the undissolved solids in the boiler after one hour?

Solution

$$\frac{SP \cdot F}{1,000,000 \; ppm} = Undissolved \; solids \; (lb)$$

Where SP = rate of steam in lb/h
F = TDS concentration of feedwater in ppm

Computation

Problem 3

$$\frac{30,000 \; lb \; steam \cdot 60 \; ppm}{1 \; hr \cdot 1,000,000 \cdot 1 \; ppm} = 1.8 \; lb \; of \; undissolved \; solids/hr$$

If the same boiler operated for seven (24 hour) days, how many pounds of solids would there be in the boiler?

Solution

$$\frac{days \cdot hours \cdot pounds \; of \; solids}{days \cdot hours} = pounds \; of \; solids$$

Computation

$$\frac{7 \; days \cdot 24 \; hr \cdot 1.8 \; lb \; of \; solids}{1 \; day \cdot 1 \; hr} = 302.4 \; lb \; of \; solids$$

At this point, if the solubility limits of the undissolved solids in the boiler water were exceeded, the solids would begin to drop out of solution, depositing scale onto the tubes in the boiler. This can cause considerable damage to the boiler and its ability to produce steam. (See Chap. 3, section entitled *"Measurement of cycles of concentration"*).

Problem 4

Say that the maximum solids in the boiler were to be held at 3000 ppm, and that the boiler feedwater and/or makeup water was 200 ppm solids.

1. What percentage of the boiler water should be blown down if the solids in the boiler water are to be maintained at a level of 3000 ppm?

2. If a steam boiler produces 52,631.6 lb of steam per hr, how many pounds of boiler feedwater will be required if the blowdown is held to 3000 ppm?

3. How many gallons of blowdown per hr will be required?

Solution 4.1

TDS concentration of boiler water   (B)   = 3000 ppm
TDS concentration of feed water   (F)   = 200 ppm
Blowdown   (PB) =   ? %

Computation 4.1

$$\frac{200 \; ppm}{3000 \; ppm} = 6.7\% \; blowdown$$

Computation 4.2

Steam loss	SL	=	52,631.6 lb/h
Blowdown in percent	PB	=	6.7 % blowdown
Blowdown in lb/h	BB	=	?

$$\frac{SP \cdot PB}{h} = \frac{FW}{h}$$

Where SP = rate of steam in lb/h
SW=rate of feedwater in lb/hr

Solution 4.2

$$\frac{52,631.6 \ lb \cdot .067}{1 \ hr} = \frac{3,526.3 \ lb}{1 \ hr} \ Feedwater$$

Solution 4.3

Rate of feedwater (FW) = 3526.3 lb/hour
Conversion factor   =   0.1198 gal/lb

$$\frac{FW}{hr} \cdot \frac{0.1198 \ gal}{1 \ lb} = \frac{gal \ of \ water}{hr}$$

Computation 4.3

$$\frac{3526.3 \ lb \cdot .1198 \ gal}{1 \ hr \cdot lb} = \frac{422.45 \ gal}{1 \ hr}$$

Problem 5

Find the percentage of condensate return using the chloride test formula (See section entitled "Testing for Cycles").

1. Measure the chlorides of the makeup water (call this chloride 1).

2. Measure the chlorides of the feedwater tank (call this chloride 2).

If chloride 1 = 20 and chloride 2 = 2,

$$\frac{2}{20} = 10\% \; chloride \; makeup \; water$$

If chloride 2 = 100% feedwater and chloride 1 = 10% makeup water, then the condensate return must equal 90%

$$PR \cdot SP = CW$$

where PR   = condensate return in percent
      SP   = rate of steam in lb/h
      CW   = rate of condensate return in lb/h

$$\frac{0.9 \cdot 49{,}105.34 \; lb}{1 \; hr} = \frac{44{,}194.80 \; lb}{1 \; hr} \; condensate \; return$$

or

$$\frac{44{,}194 \; lb \cdot 0.1198}{1 \; hr} = \frac{5{,}294.5 \; gal}{1 \; hr} \; condensate \; return$$

Makeup Water

Find the total makeup water

Feedwater to boiler (FW)	52,631.66 lb/h
Less blowdown (BB)	- 3,526.32 lb/h
Steam product (SP)	49,105.34 lb/h
Less steam loss (SW)	- 4,910.83 lb/h
Condensate return (CW)	44,194.81 lb/h

Take the feedwater to boiler and subtract
the condensate return to get the makeup
water rate (MW).

$$52,631.66 \text{ lb/h}$$
$$\underline{44,194.81 \text{ lb/h}}$$
$$8,436.85 \text{ lb/h}$$

Converting 8,436.85 lb/h to gal/h,

8,346.85 lb/h ˙ 0.1198 = 999.95 gal/h makeup water

Note: 0.1198 is a conversion factor.

Blowdown

Determine the blowdown required while keeping the solids in the
boiler at 3000 ppm (Problem 4).

Steam Product

$$(FW \text{ minus } BB) = SP$$

where SP = rate of steam in lb/h
FW = rate of feedwater in lb/h
BB = rate of blowdown in lb/h

(52,631.6 lb/h) - (3,526.32 lb/h) = 49,105.34 lb/h

Steam Loss

SP minus CW minus BB = Steam loss in lb/h (SW)

$$\frac{49,105.34 \; lb \; - \; 44,194.81 \; lb \; - \; 1 \; lb}{1 \; h} =$$

$$\frac{4,910.53 \; lb}{1 \; h} \; steamloss$$

or

SW / SP = Steam loss in percent (SL)

$$\frac{4,910.53 \; lb}{52,631.66 \; lb} = 0.093 \quad or \quad 9.3\% \; steam \; loss$$

Balancing the Equation

You have heard the saying what goes up must come down, or what goes in must come out. The same is true of all of the above equations. What goes into the boiler must come out of the boiler. The blowdown plus lost steam, plus condensate return, must all equal or balance the total amount of steam produced.

$$SP = (BB + SW + CW)$$

Blowdown	=	6.7% or	3,526.32 lb/per h
Lost Steam	=	3.3% or	4,910.53 lb/per h
Condensate	=	90.0% or	44,194.81 lb/per h
		100.0% or	52,631.66 lb/per h

52,631.66 lb/per h = 100% = 52,631.66 lb/per h feedwater rate

Therefore the equation is balanced.

CARRYOVER (FOAMING)

Carryover is the entrainment of boiler water in the steam leaving the boiler. Carryover is a very undesirable product in a steam boiler. A small amount of carryover can cause control valves in the steam system to stick, turbine blade deposits, and corrosion in steam condensate return systems. In the event of carryover where there is direct contact with food, by the steam, it can also result in food poisoning.

The reasons for carryover are many:

1. Improper boiler or feedwater design
2. Overloading the boiler or design characteristics of the boiler
3. Load surges, such as opening the boiler's steam header valve rapidly and allowing the steam to flow into the steam system very fast
4. Uneven firing of the boiler
5. High water level in steam drum
6. Poor boiler maintenance
7. Improper feedwater composition

Before carryover can be eliminated it is necessary to determine the origin, or cause, of the carryover in order to apply effective corrective measures.

The first consideration for the prevention of carryover is the separation of water from the steam when the steam leaves the boiler. To do this we must understand what happens on the surface of the boiler water as the steam leaves its liquid state and goes to the gaseous state, or steam.

When stream bubbles reach the surface of the water they perform certain actions that must be understood before carryover can be discussed (See Fig 8.5).

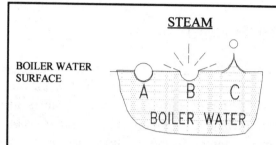

A. A steam bubble forms on the surface of the water.
B. The steam bubble bursts.
C. Boiler water rushes in to fill the void left by the burst bubble. The incoming boiler water rushes in so fast to fill the void that the water passes right on past the void and sprays small droplets of boiler water into the steam flow.

FIGURE 8.5   Carryover.

When the steam bubbles reach the water surface (A), the thin skin, forming the bubble on the surface, breaks, leaving a concave cavity in the water surface (B). The liquid then rises to fill this cavity and since the center of the void rises faster than the rest of the cavity, the velocity of the water filling the void causes a small droplet or droplets to be thrown upward (C). This is known as *misting* and is inherent in all boiler systems. It is these small droplets, *wet steam*, that must be removed from the flow of steam before it leaves the boiler.

There are several mechanical methods used for removing these droplets, such as baffles, chevron screens, centrifugal separators, and dry pipe systems. These different devices all come with the particular boiler for which they were designed.

Because of this problem, the American Boiler Manufacturers' Standards for Suspended Solids at Different Pressures were developed. (See p. 118, American Boiler Manufacturers' Standards). Water treatment companies have also developed foam dispersants (polymers) that are fed into the boiler to help prevent foaming and carryover, which work very well.

Types of Carryover

There are several types of carryover:

1. *Priming*, which is defined as the actual lifting or surging of water into the steam flow.

2. *Mist* or *fog*, which is caused by the bursting of small bubbles formed at the steam release surface.

3. *Foaming*, which results from the formation of a foam blanket on the steam release surface.

4. *Mechanical* carryover, such as silica, which volatilizes along with the water and is carried over into the steam flow. It is sometimes necessary to install a TDS sensor, in the condensate return line to detect any carryover. If the TDS sensor senses an increase in the solids, it reroutes the condensate to the drain, and at the same time adds dispersants to the boiler, until the system has cleared itself.

5. *Foreign materials*, such as oil and grease, getting into the boiler and not properly blowing down the boiler.

CHAPTER 9

STEAM BOILERS

BOILERS

The earliest boilers used at the beginning of the industrial era were simple vats or cylindrical vessels made with iron or copper plates riveted together and supported by a brick refractory. Connections on the boiler tank were made for the offtake of steam and the replenishment of water. Heat for the boiler was supplied by wood or coal. In essence, the boiler consisted of a container into which water could be fed, and by the application of heat the water would evaporate continuously into steam; something like a teakettle on the kitchen stove.

As the evolution in design and the requirements for higher boiler pressures and capacities became prevalent the designs led to the use of steel tubes, submerged in the water space of the vessel, on the inside of the boiler shell. This allowed the hot gases from the fire in the fire box to flow through the steel tubes, thus increasing the amount of heat-transfer surface. This type of boiler took the name of *fire-tube boiler* because of the heat passing through the tubes in the boiler.

Later, through evolution, the heat-transfer process was reversed. Now the tubes were filled with water and the hot gases flowed around the outside of the tubes. This type of boiler took the name *water-tube boiler*.

As time went by, boiler designs were changed to suit the environment in which they worked, such as the marine environment. Boilers were designed to supply steam to the main propulsion machinery of ships, so they were called *marine boilers*. Marine boilers became highly refined. They were usually constructed of the two-drum water-tube type, with water-cooled furnaces, superheaters, and used heat recovery equipment of the economizer or air-heater type. These boilers were first constructed in Scotland, and thus they were referred to as *Scotch marine boilers*.

Over time, as different needs for boilers came into being, other types of steam boilers were developed, such as *cast iron heating boilers*.

Cast Iron Boiler

A cast iron boiler is made up of a number of hollow cast iron sections, mechanically attached to each other. As more heat was required, more cast iron sections were added to the boiler. Hot gases, from the fire below the cast iron sections, were designed to flow around the outside of the sections heating the water and producing steam. These boilers were designed for lowpressure heating systems and were also used for hotwater boiler systems.

Fire-tube Boiler

As described above a fire-tube boiler is one in which the hot gaseous products of combustion pass through the boiler tubes which are surrounded by boiler water (See Fig. 9.1). Some boiler designs are such that the fire tubes pass through the water two and

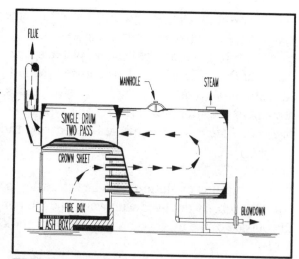

FIGURE 9.1   Fire-tube boiler.

three times before finally going to the smokestack and to the atmosphere. This allowed for more heat transfer in the boiler between the fire and the water, making the boiler more efficient. The names given to these types of boilers are *one-pass fire-tube boiler*, *two-pass fire-tube boiler* and *three-pass fire-tube boiler*. Both the water and steam in a fire-tube boiler are contained within a large-diameter drum or pressure vessel. Hence the name *shell-and-tube boiler* was also given to these types of boilers.

The shell-and-tube-type boilers are restricted to a shell diameter of 14 ft with an operating pressure of 300 lb/in² because of the stresses in the large-diameter shells and the necessity to design flat tube sheets at each end.

The heating tubes in a shell-and-tube boiler are connected to the tube sheets at each end of the boiler by flaring and rolling the ends of the tubes. These tubes also serve as structural reinforcements to help support the flat tube sheets against the force of the internal water and steam pressure. Stay bolts are also used on the inside of the boiler for structural reinforcement. If stay bolts are solid they must be 8 in or less in length. If they are threaded, they are required by the ASME boiler code to have a telltale hole drilled from the outside end of the stay bolt (center of the bolt) to a distance of about ½ in past the inside of the tube sheet. This is so that if corrosion on the inside of the boiler eats away at the stay bolt, the stay bolt will begin to leak through the hole at the open end before the stay bolt is completely gone. Leaking from the center hole in a stay bole is an indication of corrosion on the inside of the boiler.

Fire-tube boilers are best suited for low-pressure heating service, and because of their generous reserves in water capacity, they require minimum attendance. If scale is allowed to form on the water sides of the heating tubes, it insulates the heat from the fire side from reaching the water. This causes damage to the boiler tubes and lowers the heating efficiency of the boiler.

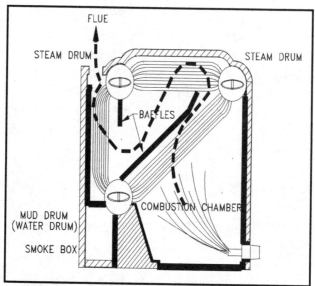

FIGURE 9.2 Two-drum water-tube boiler.

Water-tube Boiler

A water-tube boiler is one in which the water circulates within the tubes and the hot gaseous products of combustion pass over the outside of the tubes (See Fig. 9.2). One advantage of this type of boiler is that the tubes, exposed to combustion products can be small in diameter. Thus in the event of a failure of a small boiler tube, the amount of energy released is minimal so that explosion hazards of the boiler are also reduced.

Water tube boiler construction also facilitates greater boiler water and steam capacity by increasing both the number of the tubes used and their length, and by the use of steam and water drum headers. Higher boiler pressures can also be used because of the smaller-diameter tubing. Since the smaller-diameter tubes do not require an abnormal increase in wall thickness, as the pressure is increased, heat transfer is also improved.

Modern water-tube boilers generally incorporate the use of *super-heaters, economizers,* or *air heaters* to utilize more efficiently the heat from the fuel and to provide steam at a high potential for useful work in an engine or turbine.

Generators

Steam generators are defined by the American Society of Mechanical Engineers as "a combination of apparatus for producing, furnishing, or recovering heat together with apparatus for transferring the heat, so made available, to the fluid being heated and vaporized". Generators are the same as steam boilers.

The word *boiler* refers to the physical change of the contained fluid in the boiler and not to the heating process; hence, engineers prefer the term *steam generator* as the name for the primary part of a steam-generating unit.

Steam generators may include many component parts, such as the boiler, furnace, fuel-burning equipment, superheater, steam reheater, economizer, etc.

BOILER AUXILIARY EQUIPMENT

Economizer

An economizer is a tube bundle placed in the breaching or stack of the boiler for preheating feedwater. It is classified as a closed feedwater heater and utilizes waste flue products for heat energy to preheat feedwater. In effect an economizer is a water heater designed to preheat boiler feedwater and deliver the water to the steam generator or boiler at a high temperature, thus saving energy.

Deaerators

Air, oxygen, carbon dioxide, or other occluded gases carried by water may be active in producing corrosion within boilers and their associated equipment as well as in condensate return lines. These gases should be removed before they enter the boiler. Removing dissolved gases from the feedwater can best be effected by the following:

1.   Heating the feedwater going to the boiler before the water enters   the boiler. This will drive any gases in the water off to the atmosphere. (Solubility of gases in liquids decreases with an increase in the temperature of the liquid.)

2.   Increase the pressure in the feedwater tank. In a sense, squeeze the gases out of the feedwater and remove the gases to atmosphere. (Dissolved gases in a solution are directly proportional to the partial pressures of that gas in the free space above the liquid.)

Deaerators are made in several styles depending on the nature and the boiler plant operating conditions. Incoming boiler makeup water to the deaerator is broken up into a spray or film. Then steam is passed through and across the incoming water spray to force out the dissolved gases. Oxygen content, with some deaerators, can be reduced to below 0.005 cm^3/l, which is near the limit of chemical detectability.

Deaerator designs fall into two broad types, *spray deaerator* and *tray deaerator*, or combinations of the two. The typical deaerator has a heating and deaerating section plus a storage area for hot deaerated water. Deaerators operate at pressures of between 5 and 50 lb/in^2, using the incoming steam pressure from the boiler.

When looking at the discharge plume of a deaerator some engineers think the steam is being wasted, and turn the plume discharge valve off. A steam plume, coming out of the discharge pipe on the roof, is a good indication to the boiler plant operator that the deaerator is operating properly. As a rule of thumb, there should be about an 18-in plume of steam escaping at all times from the deaerator. It is always a good idea for the plant operator to check for this plume since it is a good indication of how the deaerator is operating.

FEEDWATER HEATERS

Feedwater heaters of the open or direct contact type inherently serve as deaerators. As described above, the purpose of the feedwater heater and deaerator is to drive off gases from the incoming makeup water and to heat the boiler feedwater to a high temperature so as to prevent boiler temperature strains when the boiler feedwater is first introduced into the hot water on the inside of the boiler.

The incoming feedwater is heated by internal steam coils to a temperature that is high enough to drive off the gases held in the incoming makeup water. If a boiler has a continuous skimmer type blowdown, this is sometimes used to heat the feedwater. The gases driven off by the heat escape through a vent at the top of the feedwater heater and are piped to the atmosphere.

Water supplied to a generator for the purpose of generating steam actually cools the heating surfaces of the tubes by removing heat. Steam, withdrawn from the steam drum, should always be of a high purity. Contaminants that enter the boiler with the boiler feedwater, even in small concentrations, will accumulate in the boiler water (See Introduction, section titled "Dissolved Solids").

PRETREATMENT

Water supplied to the boiler, as a rule, comes from city municipal sources, groundwater (*well*), or from surface supplies, such as lakes or rivers. Water supplied from a city or municipality is generally of good quality, but it is still unsuitable for direct introduction into a boiler because it is contaminated from contact with the earth, the atmosphere, or other sources of pollution. Whether or not *primary water treatment* is required, it will be necessary to tailor the makeup water to the needs of the boiler.

Primary water treatment could be a hot lime or lime-soda softener, a sodium zeolite softener, etc. The purpose for such a pretreatment is to reduce hardness, regulate alkalinity, and when necessary to reduce silica or total solids in the feedwater. The higher the boiler pressure, the more important it will be to have good boiler water pretreatment.

If condensate is returned to the boiler, special attention will be required to assure that the condensate is of a good quality. Oil or other contaminants should never be allowed to be fed into the boiler as they would cause *foaming*. Foaming means that when the surface of a boiler water is oily or dirty it produces a foaming action in the boiler. Foaming can lead to *priming*, which means that boiler water will *carryover* into the steam lines. Water droplets from the surface of the boiler water will carryover with the steam as it leaves the boiler and flow into the connecting steam piping system. Steam with water in it is called *wet steam*.

Feedwater to the boiler should always be virtually free of impurities, which are harmful to the operation of the boiler and its associated systems.

Boiler feedwater is usually a mixture of relatively pure condensate and treated makeup water. The proportions between the two vary with the amount of makeup water required to compensate for losses from the system served by the boiler.

Once-through or *process boilers* require very high-purity makeup water such as *demineralized* or *reverse osmosis* water.

Deposits in boilers result from the hardness contamination of the feedwater along with the corrosion products from the condensate and feedwater system and, in some cases, from the chemicals being fed to the boiler. Hardness contamination of the feedwater may result from either deficiencies in the pretreatment system or raw water contamination of the condensate.

Remember: Water is either softened before it gets into the boiler or it is softened in the boiler. Softening the water in the boiler requires chemicals to prevent scaling of the boiler.

STEAM TRAPS

Heat always flows from a higher-temperature level to a lower-temperature level. In the discussion regarding *enthalpy of steam* (See Fig 8.1) we showed that a great number of Btu were required to change the boiling point of water to steam. These Btu, or *energy,* have not been lost. They are stored in the steam, ready to be released to heat the air, cook, etc.

When the steam leaves the boiler, flowing out through the piping distribution system, the temperature of the steam is higher than the temperature of the air surrounding the steam pipes. These Btu that were stored in the steam, the *heat of vaporization* or *latent heat,* are transferred in the form of heat from the high temperature of the steam on the inside of the pipe to the lower temperature of the air on the outside of the pipe. This loss of heat, in the one case, and the acceptance of cold, in the other case, causes the steam to lose the excess Btu stored in the steam and changes the steam back to a liquid. This liquid is called *condensate.*

Using the Steam Heating Radiator as an Example

If the high-temperature steam coming from the boiler were to fill a steam radiator, the radiator would transfer the excess Btu stored in the steam to the radiator walls, heating the outside air. At the same time, on the inside of the radiator, condensate would form. Because the steam changed from a gas to a liquid condensate, the specific gravity of the condensate would be greater and the condensate would slowly fall down the walls on the inside of the radiator. Because the heat content of 1 lb of condensate is much less than 1 lb of steam, eventually the heat in the radiator would reach equilibrium with the outside air, filling the radiator with condensate. The radiator would then be unable to transfer heat since it would be cold.

To remedy this situation it is necessary to remove the condensate from the heating device as soon as it forms and replace it with hot steam, so that the radiator will continue to heat the air surrounding it.

This is the purpose of a *steam trap*. The steam trap must remove the condensate, air, and CO_2 out of the steam system as fast as it accumulates. Judgement and experience are often the most vital elements in getting the peak results from a plant's steam traps.

Condensate, caused from the loss of Btu, collects on the bottom of the piping leading back to the feedwater tank. The piping is called *condensate return piping*. Steam will continue to pass over the top of the condensate until eventually there is so much condensate in the pipe that it blocks the passage of any more steam. This causes a pressure differential between the boiler pressure and the downstream side of the steam distribution piping, which is at atmospheric pressure. The steam pressure from the boiler is higher and can push this slug of condensate water along in the piping at speeds of up to 50 to 60 mi/h. This fast-moving condensate acts like a battering ram when it comes to a turn in the pipe and can cause damage to the piping system and fittings. It can also be very annoying because of the loud sounds it makes when it hits the end of a straight piece of pipe. This action is called *water hammer*. It is essential that the condensate be removed in the form of a "heavy dew" as soon as it is formed and before it can grow into a dangerous slug of water.

Removing the condensate does not mean that the condensate is of no more value to the plant operator. Condensate is still valuable and should be returned to the boiler, if at all possible, for further use.

TYPES OF TRAPS AND THEIR OPERATION

Float and Thermostatic trap

This type of trap is used most often on applications where the condensate being drained operates on a modulating steam pressure (Fig. 9.3). That is the pressure in the heat exchanger that can vary from the maximum steam supply pressure down to vacuum, under certain conditions. Thus under zero steam pressure the pull of gravity is the only source of energy available to pull the condensate through the steam trap.

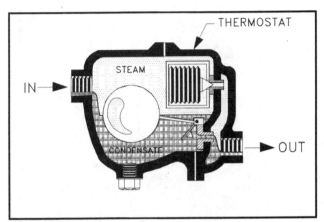

FIGURE 9.3 Float and thermostatic steam trap.

This type of steam trap is designed to remove large surges of condensate quickly, as well as noncondensable gases, such as air and carbon dioxide. The condensate in the steam trap is removed from the trap when a ball float rises. The rising level of condensate in the steam trap lifts a ball valve which is attached to a small valve, allowing the condensate to pass through the trap and return to the condensate return piping by gravity, or to be forced out of the trap by the boiler steam pressure. A thermostatic air vent, in the upper section of the trap, is responsive to temperature and will open with a change of only a few degrees. This causes any air or carbon dioxide gas to be forced out through the thermostatic vent valve. Bellows in the thermostatic air vent close as soon as the heat of the steam comes into contact with them again.

BUCKET TRAP

A bucket trap is designed to remove large amounts of condensate at a fast rate. When an initial slug of condensate, along with air and flash steam, enters the bucket trap through the trap inlet the condensate is forced down to the bottom of the trap and up through a tube on the inside of the trap by steam pressure. Above this tube is an inverted bucket, which traps any flash steam. The condensate spills over the top of the tube, down and under the bottom of the inverted bucket,

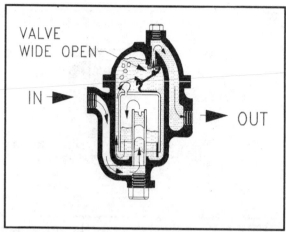

FIGURE 9.4   Inverted bucket steam trap.

and then back up to a condensate escape valve at the top of the trap. Mechanically attached to the top of the inverted bucket is a lever atop the condensate escape valve. When flash steam is captured in the inverted bucket the bucket will float to a high position, closing the condensate escape valve, preventing any steam from escaping from the trap (Fig. 9.4). If there is condensate in the trap, the bucket will drop to a low position, allowing the condensate in the trap to flow out of the trap, through the condensate escape valve, and into the condensate return line. When steam flows into the inverted bucket, the bucket rises because of the buoyancy of the steam. This closes the condensate escape valve, preventing any steam from escaping to the condensate return system. Air and carbon dioxide pass through a small vent at the top of the bucket, collecting in the top of the trap. On the next exit of condensate these trapped gases are then removed along with the flow of the condensate through the condensate escape valve.

A cast iron bucket trap can discharge about 20,000 lb of condensate per hour operating at a maximum pressure of about 250 lb/in².

To select the proper trap for a particular use it is best to call in an expert who has all of the information required to select the proper trap

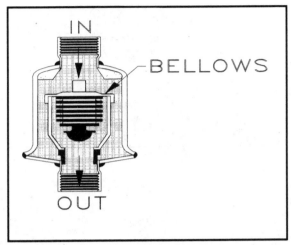

FIGURE 9.5 Thermostatic steam trap.

as well as having the knowledge and experience for a particular installation.

Thermostatic Steam Traps

This type of trap operates with heat sensitive bellows that are sensitive to heat and cold (See Fig. 9.5). When condensate builds up around the bellows, the bellows sense this lower temperature and contract, opening a needle type valve, allowing the condensate and/or air to be discharged to the condensate return line. As soon as steam reaches the bellows they sense the higher temperature of the steam and expand, closing the valve and trapping the steam.

Thermostatic traps are used extensively in cold weather, where the condensate lines might freeze. The trap is always open when water is in contact with the bellows, and the trap will completely discharge all condensate to prevent freezing.

Operating pressure for a thermostatic trap is 300 lb/in^2 and it has a maximum discharge of about 3,400 lb of condensate per hour.

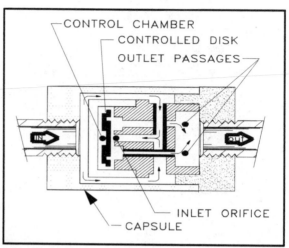

FIGURE 9.6   Controlled disk steam trap.

Disk Trap

The *disk trap*, sometimes referred to as the controlled disk steam trap, contains only one moving part, the disk itself (Fig. 9.6).

Operation of this steam trap is somewhat mysterious. Condensate and air entering the trap pass through the heating chamber around the control chamber and through the inlet orifice. This flow lifts the disk off the inlet orifice and the condensate flows through to the outlet passages. When steam reaches the disk, increased flow velocity across the face of the disk reduces pressure at this point and increases pressure in the control chamber and the disk closes the orifice. Controlled bleeding of steam from the control chamber causes the trap to open. If condensate is present it will be discharged as described above. The trap will then reclose in the presence of steam and continue to cycle at a controlled rate.

Beside the disk trap's simplicity and small size, it also offers advantages such as resistance to hydraulic shock, the complete discharge of all condensate when open, and intermittent operation for a steady purging action.

MONITORING TRAP OPERATION

There are three methods of checking the operation of a steam trap while it is operating.

Sight Method

By observing a steam trap you can determine if it is discharging hot condensate or live steam. If the trap is not discharging anything and is cool, it can be deduced that the trap is not working.

Live steam has a large number of Btu and it is hot. Live steam also discharges at a high velocity. If you see live steam coming out of the condensate return line, the steam trap has failed in the open position. If you see spurts of liquids being discharged with flash steam the trap is working properly. *Flash steam* is a relatively low-temperature vapor that forms when hot condensate is properly discharged from a high pressure to an atmospheric pressure. If the trap is not discharging live steam or condensate and it is cool, you can suspect the trap of backing up condensate: It is not working. The above methods require special piping so that the different possibilities can be observed; for this reason, these methods are not often used.

Temperature Method

A properly working steam trap should have a significant temperature differential across its body. If the trap has a hot temperature, both on the entrance and exit, it has failed in the open position, allowing steam to pass through. If the steam trap has a low temperature across both the entrance and exit, it has failed in the closed position and is probably backing up condensate.

Temperature measurements on steam traps can be made by using temperature-sensitive crayons, tapes, infrared scanners, or contact pyrometers immediately upstream and downstream of the traps and then comparing the readings.

Sound Method

By listening to a steam trap operate, it can be checked without visual observation of the discharge. Using an industrial-type stethoscope placed against the trap body, or by placing an 18 in length of metal rod at one end of the trap and the other end of the rod against your ear, you should be able to hear the trap operating. A properly working trap will operate at a fairly regular rate. A failed bucket or float trap will whistle or "whoosh" with a continuous sound of steam blowing through at a high velocity. A thermostatic trap usually fails in the closed position, so no sound at all can be heard, indicating failure of the trap. A failing disk trap will chatter rapidly.

WATER GAGES AND WATER COLUMNS

One way to know the difference between a steam boiler and a hot-water boiler when you walk into a boiler room is to look for the water gage glass on the side of the boiler. If you see a gage glass showing the level of the water in the boiler, you are looking at a steam boiler.

The water *gage glass* is the first appliance to observe on a steam boiler when checking its operation. The water gage glass shows the proper water level that must be maintained in a boiler to avoid overheating and damage to the boiler. The gage glass is usually connected directly to the water column. The water column is a hollow casting, or forging, connected by pipes at the top and bottom of the boiler's steam and water spaces. The water column is designed to give a true level of the water in the boiler by equalizing the pressure coming from the water side and the steam side of the boiler. A drain valve is located on the bottom of the water column to remove sediment, which might block the lower water side connection and thus cause a false water-level indication.

The only attachments that should be allowed on the water column are those that require no flow of water to operate, such as a pressure gage, damper regulator, pressure switch, etc.

Boiler Pressure Gage

Steam boilers are required to have at least one pressure gage, so located and of such a size that it is easily readable and which at all times indicates the boiler pressure. The gage must be connected to the steam space of the boiler, and to protect the gage from high temperature steam the gage must be connected to a siphon. A siphon is a pigtail or drop leg in the tubing to the gage for condensing steam, thus protecting the spring and other delicate parts of the gage from the high temperatures of the steam.

Safety Valve

Every boiler must have at least one safety valve, and in any case the safety valve must be capable of discharging all of the steam the boiler can generate without allowing pressure to rise in the boiler more than 6 % above the highest pressure at which any valve is set, and in no case more than 6 % above the maximum allowable working pressure of the boiler. This means that some boilers will have more than one safety valve.

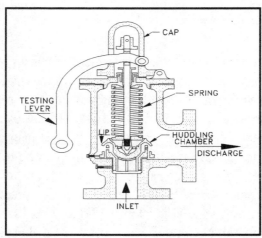

FIGURE 9.7   Pop-type safety valve.

All safety valves on a steam boiler must be of the direct spring-loaded pop-type. The pop-type safety valve has a lip or slight extension on the disk of the valve which extends beyond the seat surface and provides a huddling chamber (Fig 9.7). As the valve opens, this huddling chamber is filled with steam, thus building a static pressure on the lip because of the increased disk area exposed. This extra force upward suddenly lifts the disk against the compression spring and causes the valve to open wide almost instantaneously with a popping noise.

WATER TREATMENT

The condition of the boiler *feedwater* can be critical to both maintenance and efficiency of the boiler. Poor quality water can cause corrosion and inhibit heat transfer. However, the subject of chemical water treatment associated with steam generation is an extremely complex one and will vary greatly with the type and quality of water in each area of the country and also be dependent on the pressure and type of boiler being used.

The first step in a water treatment program is to take a water analysis of the incoming water to the system. Various chemical treatment manufacturers can be consulted and will gladly give advice on the proper chemicals to use.

Although blowdown is not water treatment, it has a great deal to do with the water treatment of the boiler. From the aspect of fuel and chemical conservation, blowdown should be kept to a minimum. There is a limit to the amount of blowdown that can be accepted without a great loss of efficiency, and there is also a limit to the amount of total dissolved solids (TDS) that can be tolerated in the boiler. These TDS tolerances will vary from plant to plant and the boiler manufacturer should be consulted. As boiler pressure rises, the maximum permissible solid level in the boiler is reduced (American Boiler Manufacturer's Standards, *boiler TDS limits*). The main objective is to keep the boiler in equilibrium, i.e., the amount of solids leaving the boiler due to blowdown should equal the amount of solids entering through the feedwater.

Dissolved oxygen (air) in the feedwater is extremely detrimental to the boiler, and its effect can be devastating as this is one of the main causes of corrosion.

Following are some of the chemicals used in boilers and their purpose.

Deposit Control

Deposits on the boiler tubes act as insulation that retards heat transfer. The insulating effect of deposits on boiler tubes prevents the steam or water from carrying away the heat. This can cause the metal temperature of the boiler tubes to rise; in turn this may cause tube failure due to the overheating of the tubes.

Internal chemical treatment for deposit control in the boiler is achieved either by adding a treatment to the makeup water, to prevent the contaminants in the boiler water from precipitating out of solution, or by adding a treatment to the boiler water that will cause the contaminants to precipitate and fall out of solution. These contaminants can then be easily removed from the boiler by blowdown.

Chelate Treatment

The Greek word *chela* means claw. Chelating tends to claw away at the undesirable calcium carbonate and calcium sulfate scales, but other anions commonly present in boiler water tend to compete with the chelant. Phosphate competes for the calcium, and if phosphates are present in significant amounts it will result in calcium phosphate precipitate. Hydroxide and silica competes with the chelant for magnesium, which could result in magnesium silicate deposits.

Chelating is one method of boiler treatment which can keep calcium hardness from precipitating out of solution in the boiler water. Chelant treatment of boiler water is attractive because the chelants of calcium and magnesium are soluble. The common chelating agents used in industrial boiler water treatment are ethylene diamine tetraacetic acid (EDTA), nitrilotriacetic acid (NAT), and/or a mixtures of the two.

Chelate treatment provides the best conditioning of hardness contaminants, but improper use can also cause corrosion in the boiler. Chelate residuals in large excess of the contaminants and chelate decomposition products can corrode boiler steel. An alkalinity residual of 200 to 500 ppm should be maintained in the boiler water at all times to minimize any potential for corrosion.

Chelate should always be fed to the boiler feedwater just after the deaerator and before the feedwater inters the boiler drum. The preferred feed location is just downstream from the boiler feed pump. A high-pressure chemical feed pump and a stainless steel injection quill are required when using chelates.

Phosphate Treatment

Phosphate is a treatment that causes contaminants in the feedwater to precipitate and fall out of solution. As a rule, phosphate is the preferred treatment for lower-pressure boilers. At low operating pressures (below 900 lb/in^2) calcium hardness, in boilers, is treated by introducing sodium phosphate to the water. Calcium introduced into the boiler water precipitates out of solution in the form of a phosphate sludge, which is removed with the boiler water blowdown before it can deposit itself on the heat-transfer surfaces.

When used in the boiler phosphates will drop out with the contaminants in the form of calcium hydroxyapatite. With magnesium contaminants in the boiler water the least troublesome precipitate is serpentine. In order for phosphate treatment to work properly it must maintain a sufficient amount of alkalinity in the boiler water in the form of hydroxides. If sufficient alkalinity is not maintained in the boiler water, where their is hardness, tricalcium phosphate will form. This precipitate tends to stick to the tube surfaces and is not easily removed by blowdown. If large amounts of phosphates are added to the boiler, the phosphate will precipitate out by itself onto the boiler tubes, causing a phosphate scale. This scale is referred to as *phosphate hideout*. On the other hand, if the phosphate residual in the boiler is not properly maintained, calcium carbonate, calcium sulfate, and/or at high pressures, calcium hydroxide scale will form on the heating surfaces of the boiler tubes.

With use of phosphate at high operating pressures (above 1500 lb/in^2), the rate of deposition is so rapid that blowdown is ineffective. In essence the most effective method for minimizing boiler water deposits at these pressures is to reduce the amount of deposit-forming material in the boiler feedwater with *pretreatment*.

A balanced amount of phosphate and hydroxide alkalinity treatment involves the maintenance of a trisodium phosphate residual with no free sodium hydroxide. Phosphate residuals are typically maintained in the range of 5 to 25 ppm PO^4, while an alkalinity residual in the boiler should be maintained in the range of 200 to 250 ppm.

Sludge Conditioners

Sludge conditioners can enhance the removal of precipitates from industrial boilers. Sludge conditioners are organic polymers that combine with precipitates and increase the electrical charges on the surfaces of the particles. This causes the particles to be dispersed and facilitates their removal by blowdown.

Insoluble Contaminants

In most applications, it is best to remove all but trace quantities of the troublesome insoluble contaminants from the feedwater before it is pumped into the boiler. Hardness in the form of calcium and magnesium salts should be removed from raw water before the water can become suitable as makeup feedwater. Metallic oxides, mainly those of iron and copper, are frequently removed from condensate returns by a filtering process. As an example, if the condensate in an underground vacuum return system is not treated properly and the pipe deteriorates, groundwater may be sucked in to the return line and contaminate the condensate. Residual trace amounts of these deposit-forming contaminants are inevitably present and additional contamination may occur from the leakage of raw water into the steam or hot side of heat exchangers (turbine condenser, for example).

Silica Contamination

Silica, in combination with other contaminants, may from an adherent and highly insulating scale. Silica vaporizes at significant and increasing rates if the boiler water temperature exceeds 500°F. If silica, in vapor solution with the steam exiting from the boiler, is not arrested by mechanical means, the resulting buildup of silica scale on equipment, such as on turbine blades, leads to reduced turbine output and costly outages. The removal of silica deposits is almost impossible and very expensive.

The concentration of silica in the boiler must be held within limits by blowdown with the maximum allowable concentration of silica decreasing as the boiler pressure increases.

Corrosion Control

When dissolved oxygen in the boiler feedwater enters the boiler, it reacts with the steel tubing and other parts of the boiler in a very detrimental way. Dissolved oxygen will travel with steam as it leaves the boiler, traveling throughout the steam lines, causing further damage to condensate return lines and steam traps. Internal corrosion in the boiler can be initiated by dissolved gases such as oxygen and carbon dioxide. Accordingly, when treating feedwater the dissolved oxygen and other gases are usually removed just before the feedwater is pumped to the boiler. Most of these gases can then be removed by boiling the water in open heaters and discharging the noncondensable gases through a vent. The most effective method of gas removal from feedwater is provided by a spray or tray type deaerating heater, arranged for countercurrent scavenging of the released gases, to prevent the gases from going back into solution in the feedwater. (See

"Deaerators" section earlier in this chapter.)

As a final polisher to the boiler feedwater, before it is pumped into the boiler the deaerator should constantly be fed with sodium sulfite at a ratio equal to the amount of feedwater being used. A 20 to 40 ppm concentration of sodium sulfite should be maintained in the feedwater at all times. The sodium sulfite will then react with any remaining oxygen, forming sodium sulfate. This will remove any final traces of oxygen before the feedwater is fed into the boiler.

Steel exposed to hot water in the boiler will corrode. Carbon steel is a much more suitable material for boiler tubes only because it forms a thin oxide film on the surface which is impervious to the water, thereby protecting the carbon steel from further attack by the water.

Corrosion Prevention

Under normal operating conditions, internal corrosion of boiler steel is prevented by maintaining the boiler water in an alkaline condition. At lower operating pressures the addition of sodium hydroxide to the boiler water will suffice to produce a pH in the range of 10.5 to 11.5. At higher operating pressures the presence of strong alkalines in the boiler water can cause metallic corrosion where the local concentration cells become established. In addition, sodium hydroxide volatilizes at high pressure sufficiently to lead to its deposition on turbine blades with the consequent reduction of turbine output. At higher operating pressures modern practice seeks to maintain only a few parts per million of sodium phosphate or a volatile amine (morpholine, cyclohexylamine, or diethylaminoethanol) in the water to keep the pH in the range of 9.0 to 10.0.

Oxygen is the most detrimental corrosion catalyst process in a steam boiler. It is responsible for pitting tubes, fire walls, and stay bolts, as well as accelerating other types of corrosion in the boiler.

For best results in protecting equipment from corrosion, a knowledgeable water treatment firm should be consulted. If the sales person does not have the answer to your problems, his or her company should be able to help.

Condensate and Feedwater System Treatment

As described above, condensate and feedwater systems are subject to corrosion by contaminants, especially oxygen and carbon dioxide gases. Oxygen and carbon dioxide gases can enter the condensate system through the leakage of air into the condensate system and/or by the release of oxygen and carbon dioxide in the boiler. Dissolved oxygen and carbon dioxide are very corrosive to both steel and copper alloys.

Bicarbonate alkalinity in the makeup water and the leakage of air may introduce carbon dioxide into the boiler system. These gases then flow with the steam from the boiler into the steam lines servicing the equipment that the steam was produced for. After the steam is used and is condensed, it is forced into steam traps. The condensate then flows back through the condensate return lines to be used again, as boiler feedwater. When carbon dioxide dissolves in water, *carbonic acid* is formed and the hydrogen ion concentration in the condensate increases. The resulting increase in the hydrogen ion concentration is very corrosive to steel and copper alloys. Corrosion due to carbon dioxide is also accelerated in the presence of oxygen, due to the depolarizing action of oxygen in the cathode reaction. By looking at a section of steel pipe, taken from a condensate return line, that has been exposed to oxygen and carbon dioxide, you can tell where the oxygen and carbonic acid have attacked the steel pipe. Oxygen attack in pipe is indicated by rust covering the inside of the pipe above the flow of the condensate. Carbonic acid, on the other hand, shows up as a groove along the bottom, and on the inside of the pipe, where the carbonic acid has attacked the steel. As a rule, where the threaded pipe joints meet, the metal is the thinnest and the carbonic acid will eat away at the threads, causing the condensate line to leak at that point.

Boiler Layup

When a steam boiler is out of service or in a standby or idle condition, it is the most vulnerable to oxygen and carbon dioxide attack. Those responsible for the operation and care of industrial and steam utility boilers must be prepared to lay up the idle equipment properly when it is removed from service, repair, or storage.

Idle boilers should be removed from service properly to reduce the tendency for suspended solids in the boiler water to adhere to the boiler metal surfaces. If a steam boiler, removed from service, contains objectionable deposits that formed during the operation of the boiler or during the improper removal of the boiler from service, chemical cleaning may be required.

Unless adequate precautions are taken, more corrosion damage can occur to the boilers and their auxiliary equipment during idle periods than entire periods of operation. When a steam boiler is removed from service for a short period of time, severe corrosion can occur very rapidly if adequate precautions are not taken. Such repeated short-term outages are frequently encountered in plants where the demand for steam is intermittent. In such situations, the need for proper layup is frequently neglected. This repeated neglect, during outage periods, can completely offset the conscientious efforts of the operator to maintain a well-run operational water treatment program. Additional protection against corrosion should be provided for deaerator heaters, closed feedwater heaters, and condensate storage tanks, during outage layup.

It is preferable to layup a boiler with complex water circuits, such as utility boilers or large industrial boilers, by the wet method.

Equipment removed from service for short-term outage, of less than seven days, is almost always laid up by the wet method, unless internal repairs are to be made or there are freezing conditions.

Functions of the chemicals used in wet layup are to keep the boiler water-oxygen free, to passivate metal surfaces, and to adjust for high pH of the water.

Specific chemicals and the concentrations at which the chemicals should be maintained in the boiler layup process should be the responsibility of the water treatment representative.

Wet Layup

The steam boiler should be completely filled with water containing the recommended concentrations of layup chemicals. This means that the boiler, if at all possible, should be filled all the way up to the steam header valve. It may be desirable to install a small gate valve just under the header valve so as to allow air from the boiler to escape when filling the boiler.

Regular tests should be made on the boiler water to be sure that the recommended levels of chemicals are maintained in the boiler.

Dry Layup

After the steam boiler has been shut down and allowed to cool, all manholes and handholes should be opened. The boiler should be thoroughly washed out, removing all dirt, rust, etc., from the inside of the boiler. The boiler should then be allowed to dry out. By dry out, I mean "bone dry." Often desiccants are used to be sure that the boiler stays dry on the inside.

CHAPTER 10

PUMPS

CLASSIFICATIONS

Pumps come in different configurations, but pumps are designed to raise, circulate, and compress liquids or gas by drawing or pressing the liquid or gas through openings and pipes.

Piston Pumps

These pumps are self-priming and work by exerting a pressure directly on the fluid being pumped. They are also given the name *positive displacement pumps*. This type of pump uses a piston with two valves on the water end of the pump. When the piston is mechanically pulled back, the first valve opens, allowing atmospheric pressure to force the liquid being pumped into the piston chamber. When the piston goes in the opposite direction the first valve closes and the second valve opens. The piston now forces the water in the piston chamber out and into the piping system. The function of discharge is merely a matter of pushing the water out of the pump. The height, head pressure, or distance it can be discharged is therefore dependent only on the strength of the pump mechanism and the power available. The same type of operation takes place in a *diaphragm pump*.

Rotary Gear Pump

The rotary gear pump is also a self-priming positive displacement pump. The unmeshing of the gears on the suction side of the pump

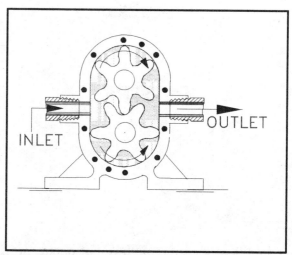

FIGURE 10.1   Rotary gear pump.

produces a partial vacuum and the atmospheric pressure forces the liquid into the pump. The liquid is then carried between the case and the gear teeth to the opposite side of the pump. The meshing of the gears forces the liquid into the outlet and into the piping system (See Fig. 10.1). Rotary gear pumps generally operate equally well when driven in either direction; however, where construction features such as built-in relief valves are involved, precautions must be taken to be sure shaft rotation is correct with respect to the special features.

The positive displacement feature of the rotary gear pump affords an exceptionally wide range of applications. This style of pump will effectively handle liquids ranging from as thin as gasoline or alcohol to as thick as heavy fuel oil or molasses.

The efficiency of rotary gear pumps depends on very close clearances between the pumping elements and the pump housing. Therefore, they are` not recommended for handling liquids with high abrasive content.

Centrifugal Pump

The centrifugal pump is a simple mechanism without valves or springs. It consists of an impeller rotating within a casing at relatively high speed. Liquid enters the impeller at the center if the impeller, or *eye of the impeller*. The impeller is designed with vanes, which sling the liquid radially outward at a velocity that is a resultant of the tip velocity of the impeller blade and the radial flow of the water through the pump. The velocity of the water is gradually decreased and the energy of motion is converted into the *pressure head* in the pump casing in which the impeller rotates. (*Pressure head* is the reading of a pressure gage at the discharge of the pump, converted to feet of liquid and referred to as datum plus velocity head at the point of gage attachment.) Centrifugal pumps can be divided into different groups classified by: *turbine and volute, number of stages, single suction and double suction, open impeller and enclosed impeller, and horizontal and vertical.*

The turbine style of centrifugal pump has an impeller that is surrounded by a casing containing diffusion vanes that are stationary. These diffusion vanes provide gradually enlarging areas, thus reducing the velocity of the liquid leaving the impeller and changing the velocity head into the pressure head. For discharge pressures greater than 50 lb/in^2, the turbine pump usually is the better pump to use.

The volute style of centrifugal pump has no diffusion vanes, but the casing of the pump is of a spiral type, forming a gradually increasing water space, thus gradually reducing the velocity of the water as it flows from the impeller to the discharge pipe. Volute pumps ordinarily have but one impeller and are considered superior to the turbine pump from a standpoint of simplicity and size, where discharge pressures are less than 50 lb/in^2.

The multistage centrifugal pump, as you would expect, consists of two or more impellers through which the water passes successively. The impellers are mounted on the same shaft and revolve in separate compartments or stages. The discharge from each stage is delivered to

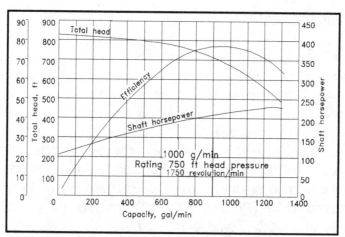

FIGURE 10.2   Typical pump characteristic curve.

the suction of the next higher stage, the final delivery pressure depending upon the number of stages. These pumps are usually of the turbine type with single inlet impellers.

For each manufacture and for each impeller, for a particular pump, the manufacturer has a *characteristic performance curve* (Fig. 10.2). The characteristic performance curve for pumps are usually based upon a constant speed of the pump. The curves show the relation between head and capacity, shaft horsepower and capacity, and mechanical efficiency and capacity. These curves are computed for a given definite speed of the pump, usually that at which the pump operates at the best efficiency or that at which it was designed to operate.

Looking at these curves, when 0 gal/min is being pumped, the total maximum head is reached, but the shaft horsepower is at its lowest. This shows that it is always best to start a large pump at what is called *shutoff*, because at shutoff, the pump draws the least amount of power. From shutoff, a valve on the discharge side of the pump can be opened slowly, allowing the pump to increase its capacity or gallons per minute as it pumps.

The quantity of liquid delivered from a pump varies directly with the speed of the impeller of the pump, the head pressure with the square of the speed, and the power with the cube of the speed. Thus doubling the speed of a pump impeller doubles the quantity of water pumped, produces a head four times as great, and requires eight times as much power to drive the pump.

SYSTEM HEAD CURVES

In addition to knowing the head for the design capacity, it is desirable to know the piping system head-capacity characteristics. When operating conditions are variable, a plot of the system head curve imposed on the pump curves enables the best pump selection for the operating range.

In any piping system the pipe friction and velocity head varies with capacity. Thus, for any fixed static head conditions, the system head increases from a static head at zero flow for any increase in capacity. Also, the static head may be variable. Then, the pipe friction and velocity head losses can be added separately to the maximum and minimum static heads, respectively, and the maximum and minimum system head curves can be plotted.

When the system curve is superimposed on the pump curve, the operating points are the intersections of the system curve with the pump curves. Thus the operating range for the pump is established and its suitability for the application is determined.

FACTORS FOR PUMP DESIGN

For all pumping design problems, it is necessary first to convert the different measurements, such as friction, pressure, specific gravity, and velocity to one type of unit, such as feet. To understand the different terms used in pumping calculations, the following terms have been outlined for your use (See Fig. 10.3).

Static Suction Lift

Suction lift, in feet, is the vertical distance from the center line of the

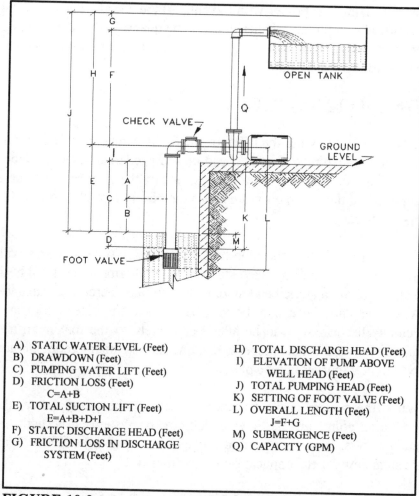

A) STATIC WATER LEVEL (Feet)
B) DRAWDOWN (Feet)
C) PUMPING WATER LIFT (Feet)
D) FRICTION LOSS (Feet)
  C=A+B
E) TOTAL SUCTION LIFT (Feet)
  E=A+B+D+I
F) STATIC DISCHARGE HEAD (Feet)
G) FRICTION LOSS IN DISCHARGE
  SYSTEM (Feet)

H) TOTAL DISCHARGE HEAD (Feet)
I) ELEVATION OF PUMP ABOVE
  WELL HEAD (Feet)
J) TOTAL PUMPING HEAD (Feet)
K) SETTING OF FOOT VALVE (Feet)
L) OVERALL LENGTH (Feet)
  J=F+G
M) SUBMERGENCE (Feet)
Q) CAPACITY (GPM)

FIGURE 10.3   Pumping system schematic.

pump down to the free level of the liquid source. *Static suction head* exists when the source of liquid supply is above the center line of the pump. Suction lift must be added to the total head in feet, while suction head must be subtracted from the total head in feet.

Sometimes suction head is given in pounds per square inch. This must be converted to feet. See the formula under *friction head.*

Static Water Level

The static water level is the distance below ground that the water is found when no pumping is occurring.

Drawdown

Drawdown is the distance that the water level drops below the static water level when the well is being pumped at its rated capacity.

Because the pump draws down the level of water when it is pumping, a drawdown distance (in feet) must be applied. The total of the drawdown distance plus the static suction lift is the equivalent of suction lift, in feet.

$$\text{Static suction lift} = \text{Pumping water level} + \text{Height of the pump above ground}$$

or

$$\text{Pumping water level} = \text{Static water level} + \text{Drawdown} + \text{the height of the pump above ground.}$$

Friction Head

This is the pump pressure required to overcome the friction loss or resistance to the flow of liquid in a piping system and it is expressed in lb/in². This friction loss can be converted into feet of liquid needed to overcome the resistance to the flow of liquid in the pipe and pipe fittings.

$$Head \ (Ft) = \frac{1 \ lb/inch^2 \cdot 2.3}{Specific \ gravity}$$

When liquid moves through a pipe it must overcome resistance caused by friction from both the liquid sliding along the pipe walls and its own turbulence. The loss in pressure caused by this is called *friction loss.* It can be calculated from friction loss tables (See Appendix B).

Static Discharge Head

The *static discharge head* is the vertical elevation from the center line of the pump to the point of free discharge.

Dynamic Suction Lift

Dynamic suction lift (DSL) is the static suction lift plus the friction losses (expressed in feet) on the suction side of the pipe.

DSL = Static Suction Lift + Friction Loss (Suction)

or

DSL = Static water level + Drawdown + Height of the pump above ground + Friction loss (suction).

Dynamic Suction Head

This includes the *suction head* minus the friction head minus the velocity head.

System Head Pressure

This is the pressure required at the discharge of the system for the system to operate properly (e.g., most domestic water systems require a system pressure of 30 lb/in^2 to operate showerheads, sprinklers, etc.).

This head pressure must be converted to feet of head pressure. See formula under *friction head.*

Total discharge Head =   Vertical height   from pump to highest point in the system + Friction losses + System Pressure (feet).

Total Dynamic Head

This includes the dynamic discharge head plus the dynamic suction lift or minus the dynamic suction head.

Velocity Head

This is the head needed to accelerate the liquid. Although the velocity head loss is a factor in figuring the dynamic head, the value is usually very small and in most cases negligible.

Submergence

This is the depth of the foot valve or submersible pump below the pumping water level. This figure does not enter into our total dynamic head calculations *except* as a pipe length when calculating friction loss.

Well Capacity

Well capacity is the maximum rate of flow a given well is able to produce without being pumped dry.

Head in Feet

Head is usually expressed in feet, whereas pressure is usually expressed in pounds per square inch. Quite often the suction lift is expressed in inches of vacuum (mercury). The formula for converting these factors follows:

Pressure

Pressure (pounds per square inch) =

$$\frac{Head \ (feet) \cdot Specific \ gravity}{2.31}$$

Head (feet)

$$Head\ (feet) = \frac{Pressure\ (lb/inch^2)\ \cdot\ 2.31}{Specific\ gravity}$$

Vacuum

Vacuum (inches of mercury) =

$$DSL\ (feet)\ \cdot\ 0.883\ \cdot\ Specific\ gravity$$

Specific Gravity

Specific gravity is the direct ratio of any liquid's weight to the weight of water at 62°F. Water at 62°F weighs 8.33lb/gal and is designated as 1.0 sp. gravity.

Viscosity

Viscosity is the property of a liquid that resists any force tending to produce flow. It is the evidence of cohesion between the particles of a fluid that causes a liquid to offer resistance analogous to friction. An increase in the temperature reduces the viscosity; conversely, a temperature reduction increases the viscosity. Pipe friction loss increases as viscosity increases.

Friction

Friction head is the pressure (in terms of feet of liquid) required to overcome the resistance to flow in pipe and fittings. Friction of liquids in pipes increases as the square of the velocity.

The size of pipe and pipe fittings for any installation should be large enough to keep friction losses reasonably low. The velocity should be kept within 10 f/s for good practical results.

All charts for determining friction losses are based on these losses for clean steel pipe on schedule 40 and show average values for new pipe including an adjustment of 15 percent for commercial installations. To obtain approximate values for other types of pipe, use multiplier correction factors of 0.9 for smooth pipe and 1.43 for 15-year-old pipe. These are rough estimates. Valves and fittings should always be taken into account when calculating pipe friction losses. Friction loss in valves and fittings are converted to feet of pipe, and tables will tell how much friction loss there is at different velocities.

If the total dynamic suction lift (or head) and the total dynamic discharge head have not been figured, list them separately. List the suction and discharge lengths, the size of piping including all fittings, static elevations, and pressure required at the discharge nozzle, if any.

Many pumps are installed as replacements where it is necessary to utilize existing suction lines. In such places, the lift and friction of the piping should be carefully analyzed so that the pump selected will be one that does not exceed the pipe capacity.

Suction piping should be as short and direct as possible. It should be the full size called for by the pump, and if the line is long, the size should be increased. All horizontal runs should slope up to the pump so as to avoid air pockets in the suction line.

Cameron's Hydraulic Data (Ingersoll-Rand Co., Cameron Pump Division, Woodcliff Lake, N.J. 07675. 1970) and the *Hydraulic Hand Book* (Fairbanks Morse Pump Division, Kansas City, Kansas, 66109) are good source reference books.

EXAMPLE FOR SIZING A PUMP

Facts

1.   Required: 800 gal/min at 70 lb/in^2
2.   Suction head pressure of 60 lb/in^2 (city water pressure)
3.   Static head of 639.80 ft
4.   800 ft of 6 in pipe
5.   three 6 in gate valves
6.   one 6 in check valve
7.   ten 90° long sweep elbows

Problems

1.   What size centrifugal pump is required?
2.   What horsepower motor for the pump is required?

Computations

Suction head pressure

$$SHP = \frac{60 \ lb/inch^2 \cdot 2.31}{Specific \ gravity \ of \ 1} = \qquad - 138.60 \ ft$$

Discharge head pressure

$$DP = \frac{70 \ lb/inch^2 \cdot 2.31}{Specific \ gravity \ of \ 1} = \qquad + 161.70 \ ft$$

Use the pressure drop of water through schedule 40 steel pipe and resistance of valves and fittings to flow of fluids (See Appendix B) .

$$Pipe = 800 \; ft \; (6 \; inch \; pipe) \cdot \frac{1.78 \; LB/inch^2}{100 \; ft} = 14.24 \; ft$$

$$Gate \; valves = 400 \; ft \cdot 3 \; valves \cdot \frac{1.78 \; lb/inch^2}{100 \; ft} = 24.48 \; ft$$

$$Checkvalve = \frac{40 \; ft \cdot 1 \; ck \; valve \cdot 1.78}{100 \; ft} = 0.71 \; ft$$

$$Long \; sweep \; elbows = \frac{12 \; ft \cdot 10 \cdot 1 \cdot 1.78}{100'} = 2.31 \; ft$$

Adding the above answers gives a total friction loss, in feet of head:

Suction pressure	- 138.60 ft
Discharge pressure	+161.70 ft
Static head	+639.80 ft
Pipe friction loss	+ 14.24 ft
Gate valves friction loss	+ 21.36 ft
Check valve friction loss	+ 0.71 ft
Elbows friction loss	+ 2.31 ft
Total head	**+701.52 ft**

System Head Curve

Imposing the 800 gal/min and the 701.52 ft head that was calculated above on the pump characteristic curve (Fig. 10.2), we get results as shown on the system design curve (Fig.10.4). From all indications, this would be the right pump for the above problem.

Solution

1.   A pump that will pump 800 gal/min at 701.52 ft head pressure will be needed.

2.   The brake horsepower (*shaft horsepower*) is indicated at about 200 hp in the pump chart (Fig. 10.4).

The brake horse power (BHP) can also be calculated as follows:

$$BHP = \frac{Rate\ (gal/\text{min}) \cdot Total\ head\ (ft)}{3960 \cdot Pump\ Efficiency}$$

$$\frac{800\ gal/\text{min} \cdot 701.52\ ft\ head}{3960 \cdot 0.7} = 202.54\ BHP$$

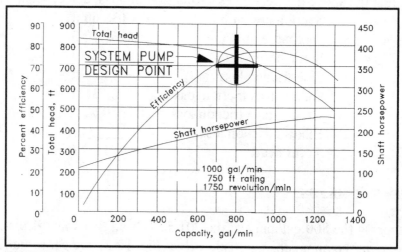

FIGURE 10.4   System design curve.

NET POSITIVE SUCTION HEAD

(*Net positive suction head*) is one of the most confusing terms in the pumping industry. It is a major problem encountered in many pumping applications, particularly those involving fluids at or near their boiling points and it is used very often in pumping applications such as boiler feedwater pumps, where the liquid being pumped is very hot. An analogy of NPSH is that in years past, automobile engines sometimes would stop running during hot weather because the fuel coming from the gas tank would get so hot it would vaporize into a gas and the fuel pump would be vapor-locked. This prevented the fuel pump from operating and gas could no longer get to the carburetor.

One must understand that pumps *do not suck liquid.* The term *suction lift* is misleading because no pump can lift liquid. Liquid is pushed into the pump by atmospheric pressure . The way a pump gets liquid up into the suction pipe is to draw the air out of the suction pipe. As the air pressure in the suction pipe is reduced, atmospheric pressure pushes the liquid into the suction piping. The pump gives the illusion that it can lift a liquid.

The relationship of pressure to height of water in a suction pipe is that a column of water 2.31 ft   high exerts a pressure of 1 lb/in^2. Atmospheric pressure at sea level is 14.7 lb/in^2 and will therefore support a column of water 33.9 ft high if there were a perfect vacuum in the suction pipe.

$$\frac{14.7 \ lb/inch^2 \ \cdot \ 2.31 \ ft}{1 \ lb/inch^2} = 33.9 \ ft$$

Pumps are not that efficient and atmospheric pressure also reduces with altitude (See Table 10.1).

It is important to know that liquid will boil at any temperature if the pressure is reduced sufficiently. The inverse to that rule is that the

higher the pressure on the surface of a liquid, the higher the boiling
point of that liquid. It is the problem of the system designers to make
certain that there is a *sufficient pressure on the fluid being fed to the
pump* so that the liquid does not boil in the suction system of the pump.

The reason that a pump requires positive suction is that a pressure drop
occurs between the pump suction flange and the minimum pressure point
within the pump impeller (See Fig. 10.5) because of the following:

1. An increase in the velocity between the suction flange and
   entrance of the impeller vanes

2. Friction and turbulence losses between the suction flange and
   entrance to the impeller vanes

TABLE 10.1

Altitude vs. Atmospheric Pressure	
Altitude above sea level, ft	Atmospheric pressure at 75°F, ft of water
0	34.0
500	33.4
1000	32.8
1500	32.2
2000	31.6
2500	31.0
3000	30.5
3500	29.9
4000	29.4
4500	28.8
5000	28.3
5500	27.8
6000	27.3
6500	26.7
7000	26.2
7500	25.7
8000	25.2
8500	24.8
9000	24.3
9500	23.8
10000	23.4

When a liquid flows through the suction line and enters the eye of the pump impeller an increase in velocity of the liquid takes place. This increase in velocity, of course, is accompanied by a reduction in pressure. If the pressure falls below the vapor pressure corresponding to the temperature of the liquid, the liquid will vaporize and the flowing stream will then consist of liquid plus pockets of vapor. Continuing with this stream flowing through the impeller, the liquid reaches a region of higher pressure and the cavities of vapor collapse. It is this collapse of vapor pockets that causes the noise coincident with cavitation. *Cavitation is the formation and collapse of low-pressure vapor cavities in a flowing liquid.*

A pump can operate rather noiselessly yet be cavitating mildly. The severe cavitation will be very noisy and will destroy the pump impeller and/or other parts of the pump.

All pump systems must have a positive (higher) suction pressure,

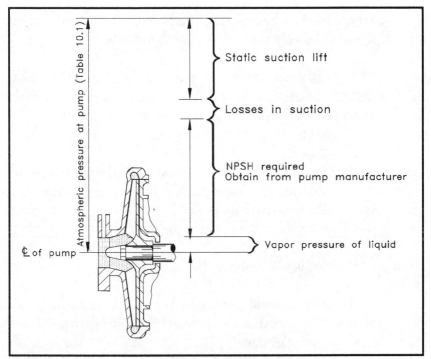

FIGURE 10.5   Centrifugal pump section.

sufficiently high enough to overcome this pressure drop within the pump and to keep the fluid from boiling and turning into a gas at the eye of the pump impeller. For centrifugal pumps avoid as much as possible the following conditions:

1. Heads lower   than head at peak efficiency of pump.

2. Capacity much higher than capacity at peak efficiency of pump.

3. Suction lift higher or positive head lower than recommended by manufacturer.

4. Liquid temperatures higher than that for which the system was originally designed.

5. Speeds higher than manufacturer's recommendation.

NPSH can be defined as the head that causes liquid to flow through the suction piping and finally enter the eye of the impeller.

Required NPSH is a function of the pump design. It varies among different makes of pumps, among different pumps of the same make, and with the capacity and speed of any one pump. The required NPSH is a value that must be supplied by the maker of the pump.

Available NPSH is a function of the system in which the pump operates. It can be calculated for any installation. Any pump installation, to operate successfully, must have an available NPSH equal to or greater than the required NPSH of the pump at the desired pump conditions.

When the source of liquid is above the pump:

NPSH = Barometric pressure (ft) + Static head on suction (ft) - Friction losses in suction piping (ft) - Vapor pressure of liquid (ft)

$$H_{sv} = Hvp + Hz - Hf - Hp$$

<u>When the source of liquid is below the pump</u>:

NPSH = Barometric pressure (ft) - Static head on suction (ft) - friction losses in suction piping (ft) - Vapor pressure of liquid (ft)

$$Hsv = Pa + Hz - Hf - Hvp$$

Where Hsv = NPSH expressed in feet of water.
Hp = absolute pressure on the surface of the liquid where the pump takes suction expressed if feet of water.
Hvp = absolute vapor pressure of fluid at the pumping temperature expressed in feet of water
Hz = static elevation of the liquid above the centerline of the pump expressed in feet of water. If the liquid level is below the pump centerline, Hz is minus.
Hf = friction and entrance head losses in the suction piping expressed in feet of water.

Problem

Let us assume that a boiler feedwater pump needs to be replaced. What requirements should be given to the pump salesperson regarding available NPSH for this particular installation?

Facts

1. The pump sits at an elevation of 5000 ft.
2. The boiler is operating at 15 lb/in^2.
3. The existing pump was designed for 30 gal/min. (Note: A boiler feed pump should always develop a pressure higher than the boiler pressure.)
4. The feedwater temperature is 200°F at 5000 ft with a vapor pressure of 26.6 ft.

5.   The static elevation of the liquid above the pump is
      72 in, (note inches must be converted to ft).
6.   There is one 2-in gate valve, one 2-in, 90°-elbow leading
      to the suction of the pump and 4 ft of 2-in pipe.
7.   The pump sits below the feedwater tank.

Computation

For this application:

$$Hp = \text{Vapor pressure of fluid in ft} = Hvp$$

$$Hsv = +Hvp + Hz - Hf - Hvp$$
$$\text{cancel}$$
$$Hsv = \quad + Hz - Hf$$

Hz = 72-in above the eye of the impeller. Because all
calculations must be in feet.

$$Hz = \frac{72 \ inch}{12 \ inch} = 6 \ ft$$

Hf. = 0.09 ft

Gate valve

$$\frac{1.3 \ ft \cdot 0.804 \ lb/inch^2}{100 \ ft} = 0.01 \ lb/inch^2$$

90° Elbow

$$\frac{3.5 \ ft \cdot 0.804 \ lb/inch^2}{100 \ ft} = 0.06 \ lb/inch^2$$

4 ft of 2-in pipe

$$\frac{4\ ft \cdot 0.804\ lb/inch^2}{100\ ft} = 0.06\ lb/inch^2$$

By adding the above totals we will know the total friction loss.

One 2-in gate valve	=	0.01 lb/in²
One 2-in 90° elbow	=	0.02 lb/in²
Four ft of 2-in pipe	=	0.06 lb/in²
Total	=	0.09 lb/in²

Converting lb/in² to ft of head

$$Hf = 0.09\ lb/inch^2 \cdot 2.31 = 0.2\ ft$$

ANSWER

$$Hsv = 6\ ft - 0.2\ ft = 5.8\ ft$$

The pump person should be told that the pump requirements are:

25 lb/in²
30 gal/min

and that there is an available NPSH of 5.8 ft

CHEMICAL METERING PUMP

There are two main types of chemical metering pumps. One is the *peristaltic* type of pump. This is a positive displacement type pump that has a plastic tube wrapped around a shaft with rollers on it. As the shaft turns, the rollers squeeze the plastic tubing, forcing chemical out of the tubing under pressure. The limiting of the pressure is the plastic tube, which is constantly flexed, as the rollers roll over it. The maximum

operating pressure for this type of pump is about 25 lb/in². The pump
will pump variable amounts of chemical from 2 to 10 gal/day.

The other type of pump is called a *diaphragm pump* and it is also a
positive displacement pump. This pump can use an electric solenoid
or an electric motor drive. The advantages of the electric solenoid type
of pump is that it can be sealed watertight, so that the pump
environment is not too important. The electric motor type requires air
to cool the motor gears, etc., and has to be operated in a cleaner

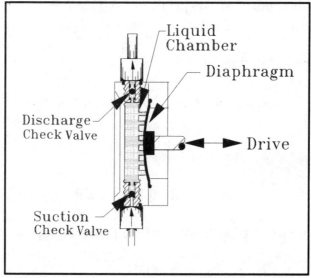

FIGURE 10.6   Sectional drawing of a positive
displacement chemical pump head.

environment. Both types have been greatly refined to the ultimate and
are good operating pumps. Depending on the drive, the pumps can
operate at pressures up to 250 lb/in².   These pumps can pump
amounts of chemical depending on viscosity of the liquid from 2 to 250
gal/day.

The liquid end of both of these pumps operate in the same way. They
are diaphragm type pumps, where the diaphragm moves in and out, both
sucking in liquid and forcing it out at the top of the pump. The only

difference is the type of valving and tubing connections, which have been perfected by the different manufacturers for their pumps.

These types of pumps can pull a good suction of several feet, but in all cases, the suction fittings must be airtight. One of the biggest problems in using this type of pump is the fittings. All of the suction fittings must be tight.

A suction valve is also required for the pumps to operate properly. The suction valve should have a strainer, so as to prevent any dirt particles from entering the pump. Incorporated into the suction valve is a movable ball, similar to a ball bearing. This ball sits on an O ring, making a tight seal, and prevents any liquid from leaking back to the pumping liquid. If the ball does not sit properly and liquid leaks back out of the suction hose, air has to replace the liquid leaking out. So where does the air come from? It would have to come from the fittings. They must be airtight.

As the pump diaphragm operates, atmospheric pressure pushes the liquid into the suction valve, lifting the ball check valve, which allows the liquid to follow the suction piping up and into the pump.

Looking at Fig. 10.6 there are three parts to the pump: (1) suction valve, (2) diaphragm, (3) discharge valve. When the diaphragm is pulled back, it creates a vacuum. This vacuum pulls down on the discharge valve, making a tight seal, and pulls up on the suction valve, allowing fluid to enter the diaphragm chamber.

When the diaphragm is pushed forward, it does just the opposite of the above. It pushes down on the suction valve, making a tight seal, and pushes up on the discharge valve, allowing the liquid in the diaphragm chamber to flow out of the discharge valve.

It is the force of the pressure on the diaphragm that determines the pressure of the pump. It is the volume in the diaphragm chamber that determines the volume of the pump.

CHAPTER 11

PRINCIPLES OF
ION EXCHANGE SOFTENING

DEMINERALIZERS AND SOFTENERS

Demineralized water is often used in process boilers because the solids have been removed and will not contribute to the scaling of the boiler.

The process of demineralization is the same type of process used in a water softener, except that in the ion exchange process all of the ions, both negatively charged and positively charged, are removed. Materials capable of attracting both of these ions, positively charged cations and negatively charged anions are required. Sometimes these materials are mixed together, called *mixed-bed demineralizers*, and sometimes they have two separate tanks, one for removal of the positive ions and one for the removal of negatively charged anions. These are called *two-bed demineralizers.*
When the resins are exhausted, they must be regenerated with a strong acid and a strong base to restore their ion exchange capacity. Cation resin is typically regenerated with hydrochloric or sulfuric acid. Anion resin is normally regenerated with sodium hydroxide.

The quality or degree of demineralization is generally expressed in terms of specific resistance (ohms or specific conductance, mhos) (see Chap. 3, section entitled "Electrical Conductivity"). Ionized material in water conducts electricity. The more ions, the more conductivity and the less resistance. When ions are removed, resistance goes up and therefore the water quality has been improved. The more scaling material that is removed, the better the quality of the water.

Silica removal and carbon dioxide removal are usually accomplished by the use of strong base resins. Mixed-bed unit pH is typically around 7.0 before the water is exposed to the atmosphere, because of the almost complete demineralization that occurs.

It must be remembered that when all of the ions are removed from the water, the water becomes very aggressive and will want to attack metals. Demineralized water is very corrosive, and when used in a steam boiler the water in the boiler must be treated to protect the boiler.

ION EXCHANGE PROCESS

Softening of water by the ion exchange process involves the exchange or substitution of the hardness in minerals that are found in water-chiefly the exchange of calcium and magnesium minerals for sodium minerals. The exchange is made possible because the minerals are ionic in nature often called *ionized impurities* which means they have an electrical charge. The ions can be either positive or negatively charged with electrons. The positive ions are called *cations* and the negative ions are called *anions*. It is these positive cations in the form of calcium, magnesium, iron, and manganese that cause the hardness that is associated with water. Removal of these hardness ions via ion exchange is the process used for softening water.

The ion exchange process is based on the fact that like charges repel one another and unlike charges attract.

Water is commonly referred to as the *universal solvent,* and when it trickles down through the strata of rock and soil it dissolves the calcium and magnesium deposits in the earth. This dissolved rock eventually finds its way into an underground aquifer. When water from the aquifer is pumped to the surface it contains the dissolved hardness minerals or calcium and magnesium, which is called *hard water*.

Hard water can cause scaling in boilers and in household appliances. It will produce a curd with soap before it will produce a lather, thus soiling kitchen and bathroom fittings, giving laundry a less clean appearance and wasting soap.

The ion exchange method of softening has become a widespread treatment process: the original and synthetic type zeolite materials have been replaced over the years by newer and more versatile ion exchangers, such as the sulfonated coals, the sulfonated phenol formaldehyde type resins, and more recently by the physically and chemically resistant high-capacity materials of the sulfonated polystyrene bead type zeolite.

The softening of water by ion exchange relies on the replacement of the calcium and magnesium ions in the water by an equivalent number of sodium ions.

As the complex raw water enters the exchange tank the positive hardness ions exchange in the resin, displacing the positive sodium ions which were in the resin to the service stream as soft water. Because calcium and magnesium are positive cations, they will repel other positive cations. The resin, being charged with positive sodium ions, exchange with the positive calcium and magnesium ions. This is because all cations do not have the same positive charge strength. Calcium and magnesium now occupy the exchange sites on the resin beads and the displaced sodium cations pass downward through the resin "bed" and out of the softener, delivering "soft" water. This process will continue for some time until eventually all of the resin exchange sites on the resin are occupied by calcium and magnesium and no further exchange can take place in the softener. When this occurs, the hardness ions begin to leak out the bottom of the resin bed. For all practical purposes, the resin is exhausted with calcium and magnesium and has no more sodium available to displace. It is at this point when the softener must be regenerated.

The resin, polystyrene divinylbenzene, which is used in most softeners, consists of millions of tiny plastic balls (beads), all of which contain many negatively charged exchange sites which attract positive cations. The resin itself prefers the more strongly positively charged cations, such as calcium and magnesium, rather than the weaker positively charged sodium cations. Thus when a positively charged calcium molecule comes by, the resin gives up the sodium cation and attracts the positively charged calcium cation: It makes an exchange.

Ion exchange is possible for two reasons: (1) All cations do not have the same strength of positive charge. (2) The negatively charged resin prefers the more strongly positively charged cations, such as calcium and magnesium, than the weaker positive sodium cations.

The resin in the softener is regenerated with a solution of sodium chloride (common salt) mixed with water, which is called *brine.* During regeneration of the softener, the flow of service water from the softener is stopped. Brine is then drawn from a separate brine tank, mixing with a separate stream of hard water. This brine solution then flows downward through the resin in the softener tank, contacting the resin beads, which are loaded with calcium and magnesium ions. Even though the calcium and magnesium ions are more strongly charged than the sodium ions, the concentrated brine solution contains literally billions of the more weakly charged sodium ions, which have the power to displace the smaller number of calcium and magnesium ions which were deposited on the resin. As the calcium and magnesium ions are displaced (exchanged), they are washed down the drain. The positive sodium ions are then attracted to the negative resin exchange sites. Eventually, all of the resin exchange sites are taken up by the positive sodium ions and the resin is said to be regenerated and ready for the next softening cycle (Fig. 11.1).

ION EXCHANGE CYCLE

During regeneration of the softener, the flow of soft service water from the softener is first stopped. The regeneration process then utilizes a five-cycle control process which consists of the following: (1) backwash, (2) brining, (3) slow rinse, (4) fast rinse, and (5) brine refill.

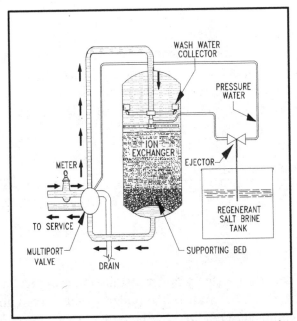

FIGURE 11.1 Single-bed ion exchange unit.

The resin is first backwashed (moves water up-flow through the resin bed) to remove traces of finely divided material that may be present and to classify the particles (coarse material at the bottom and finer material at the top of the bed).

After the particles of resin settle, following the backwashing, the regenerant is passed through the resin bed. The *regenerant* is a solution of sodium chloride (common salt) mixed with water, which is called *brine*.

The resin bed is then rinsed with a slow rinse and then a fast rinse until the excess and waste regenerant has been removed. At this time the water to be treated or softened is passed through the resin until the capacity of the resin is again exhausted.

Hardness

Hardness is usually expressed in engineering terminology as *grains* of hardness per gallon (as calcium carbonate). Sometimes it is expressed as *ppm hardness*. One grain per gallon equals 17.1 parts per million. To convert parts per million to grains the following formula can be used:

$$\frac{Parts \; per \; million \cdot grains}{17.1 \; parts \; per \; million \cdot gallons} = Grains \; per \; gallon$$

Sizing

Sizing a softener for a steam boiler installation is dependent on three things: (1) the peak flow rate of the boiler, or the maximum flow rate (in gallons per minute) required by the boiler at any given time during the day; (2) the total gallons of makeup water required by the boiler per day; and (3) the total grains of hardness per gallon in the makeup water.

Peak Flow Rate

This can be determined by using the boiler horsepower or by knowing the number of pounds of steam per hour that the boiler produces. Sometimes boilers are operated at twice their rated capacity, so as a safety margin we use twice their rated capacity.

Using Boiler Horsepower

Peak flow rate per minute = (Boiler) · (4 gal per hour) · (total hours of boiler operation) - (percent of condensate return) + (daily gallons of blowdown / hours of operation) · 2 / 60 min / h.

Using Pounds of Steam

Peak flow rate per minute = (pounds of steam per hour / 8.345 pounds per gal;) · (total hours boiler operation) – (percent of condensate return) = (daily gallons of blowdown / hours of operation) · 2 / 60 min / h.

Water Usage

To determine the number of gallons used per day by the boiler take the peak flow rate per minute and multiply it 60 minutes per hour, and then multiply that by the number of hours the boiler is in operation.

Hardness Capacity per Day

To determine the total grains of hardness per day for a water softener, multiply the water usage by the gallons of hardness in the boiler makeup water.

WATER SOFTENER CALCULATIONS

Problem

How many grains per day of water softener would be required for a steam boiler with the following characteristics when the hardness of the makeup water is 12 grains per gallon?

Given: 300 boiler hp operating 18 hours per day with 30 percent condensate return.

Computation

Makeup water:

$$\frac{300 \ hp \cdot 4 \ gal}{1 \ h \cdot 1 \ hp} = \frac{1200 \ gal}{1 \ h} \quad Makeup \ water$$

Condensate:

$$\frac{1200 \ gal \cdot 0.3 \ condensate \ return}{1 \ h} = \frac{360 \ gal}{h} \ Condensate$$

Blowdown:

$$\frac{40 \ gal \ blowdown \cdot day}{1 \ day \cdot 18 \ h} = \frac{2.22 \ gal}{h} \ Blowdown$$

Gallons of boiler water required per day:

$$(Makeup \ water) - (Condensate \ return) + (\frac{blowdown}{1 \ h}) \cdot 2$$

$$= \frac{gal \ water}{day} \ required$$

$$(\frac{1200 \ gal}{1 \ h}) - (\frac{360 \ gal}{1 \ h}) + (\frac{2.22 \ gal}{1 \ h}) \cdot 2$$

$$= \frac{1684.44 \ gal \ water}{day} \ required$$

Grains required per day:

$$= \frac{1684.44 \ gal \ water}{1 \ day} \cdot \frac{12 \ grains}{1 \ gal} = 497,592 \ grains \ Softener$$

ANSWER

It would require a softener capable of supplying 497,592 grains in one day.

SALT CONSUMPTION

Two factors must be taken into consideration when sizing a modular system for capacity: (1) Salt or chemical dosage, and (2) Number of regenerations per day.

Salt of chemical dosage: The amount of salt or chemical used to regenerate a cubic foot of resin or mineral determines operating cost as opposed to initial cost.

Salt dosage: To obtain maximum efficiency it is recommended to use an optimum salt dosage of 5 lb/ft^3 of resin to maintain the lowest operating cost. Check for the type of resin in your softener to see how many grains your resin will deliver per cubic foot of resin when using one pound of salt. Check with your water softener supplier to see how many cubic feet of resin your softener uses.

To obtain maximum capacity use the maximum salt dosage of 15 lb/ft^3 to obtain the maximum capacity and a slightly lower initial cost. A water softener supplier will have the salt dosage vs. capacity for the resin that is used and will be glad to supply you with this information (See Table 11.1).

NUMBER OF REGENERATIONS PER DAY

The capacity of a given piece of equipment for water softener can be doubled simply by regenerating it twice a day, and quadrupled by regenerating it four times a day. Start by regenerating the softener once per day. By checking the hardness of the water, it can be determined if the softener needs to be regenerated more often.

There are many ways to control the number of regenerations. Use of a time clock is probably the most common. Then there are water meters and electronics. But for all the timing devices, a time clock is still used to sequence the operations of the water softener.

NOTE:

Saturated brine is when salt dissolves in water to 25 percent.

> 1 gal of 26% brine has 2.6 lb of salt.
> 1 gal of 26% brine solution weighs 10 lbs.
> 1 ft^3 of 26% brine has 19.5 lb. of salt.
> 1 ft^3 of 26% brine solution weighs 75 lb.

Table 11.1

Brine tank capacity and area chart

Tank diameter, in	Tank area, ft^3	Inches of height, gal	Saturated brine solution, lb
18	1.67	1.10	2.86
20	2.67	1.33	3.48
24	3.14	1.95	5.07
30	4.90	3.04	7.90
36	7.06	4.40	11.40
42	9.26	5.97	15.50
48	12.57	7.80	20.20
54	15.90	9.90	25.20
60	19.63	12.20	31.80
66	23.76	14.70	38.20
72	28.27	17.50	45.50

REVERSE OSMOSIS

Reverse osmosis is the reverse of the natural osmosis process that occurs in nature. Osmosis is the natural passage of a liquid through a semipermeable membrane, during which the liquid flows from a state of low concentration of solids or impurities to a state of relatively high concentration. To treat boiler water by the osmotic process, osmosis must be reversed to make the water flow from a state of high concentration of solids to a low concentration. This is done by applying pressure to the high concentration side of the membrane; the result is a high quality water. The reverse osmosis process has proved to be a highly efficient, low-cost method of obtaining high quality water.

The semipermeable membrane must be made of a highly resistant material since it must withstand pressure higher than the somatic pressure, which can be very high.

The membrane made from cellulose acetate or polyamid must be constructed in a functionally suitable form, termed a *module*. In addition, organic matter as well as colloids and turbidity are almost 100 percent removed from the treated water.

OPERATION

The prefiltered water for treatment is brought up to 200 lb/in^3 by means of a pressure pump. In the industrial series pressures can go as high as 400 lb/in^2. This is greater than the osmotic pressure of the water. The "permeate"(pure water) transfers through the membrane while the salts and other dissolved particles are piped off to the drain with the remaining part of the water, the "concentrate" (waste water).

The correct choice of pretreatment is very important as it influences the quality and quantity of the permeate, and above all, the life-span of the module.

With increasing operating pressure and temperature of the feedwater, the capacity of the reverse osmosis (RO) installation increased also. The capacity of the installation must be measured against the lowest feedwater temperature. Since the rated output is based on feedwater temperature of 25° C (77° F), it must be divided by the appropriate temperature correction factor. Fouling can cause the capacity of the module to be reduced by about 20 percent at 400 lb/in^2 and by 8 percent at 200 lb/in^2 in three years.

There are different theories about the water and salt transfer mechanism through the membranes. One assumption is that capillaries are situated in the surface of the membrane through which the pure water flows. Another refers to a solution-diffusion phenomenon. In both cases dissolved inorganic and organic pure water is forced through the membrane and dissolved impurities remain on the feed side of the membrane. The recovery rate of permeate can amount to 35 to 50 percent in cellulose-acetate installations.

FILTER TYPES

Air quality for cooling towers at any particular site can be the cause of serious adverse effect upon the water quality. This is because cooling towers are extremely effective air washers.

The constant washing of the incoming air by the cooling tower plus the base characteristics of the makeup water supply are the parameters which establish the ultimate quality of the recirculating tower water. This is complicated by the fact that in the process of evaporation, the incoming contaminant levels of total dissolved solids concentrate tremendously.

Suspended solids that are brought into the tower system water from the atmosphere can build up to several inches deep at the bottom of the cooling tower due to the washing of the incoming air. This bed of mud, in the bottom of the tower, is one of the best breeding grounds for microbes and other disease-causing bacteria. To reduce the cost of biocide, corrosion, and the fouling of condenser tubes, this mud must be removed from the tower.

> *The two best ways of removing mud or silt is with a sand filter and a solid separator.*

Both of these systems when installed are designed so that some of the tower water, in the case of a sand filter, or all of the tower circulated water will flow through them. These systems continuously remove the suspended solids which are in the water, and then the solids are backwashed to drain.

Both of the above systems will help to prevent severe fouling of the cooling tower system, provide less energy consumption, and reduce the amount of chemicals required for maintaining the tower water in good operating condition.

Suspended matter that is found in water can often be removed by means of mechanical filters. Some of the more common types of mechanical filters are cartridge filters, sand filters, and solid separators.

Cartridge-type Filters

Cartridge-type filters are best suited for closed systems because closed system water has mill scale, corrosion particles, and other contaminates. For good heat transfer efficiency and maintenance this water should be kept as clean as possible. To do this it is best to install a sidestream cartridge-type filter. Cartridge-type filters are designed to collect the suspended solids found in the water on the outside of the filter cartridge or on the largest surface area of the filter. Filter cartridges are usually about 9-3/4 in high · 2-3/4 inches in diameter with a 1-in hole in the center. The cartridge fits over a mandrel that is on the inside of the cartridge holder. Incoming water then flows around the outside of the cartridge filter, through the filter, filtering the solids as it goes, and then to the center hole of the cartridge where the clean water is directed out of the cartridge holder.

Materials used for manufacturing the filter cartridge are determined mainly by the type of suspended solids and the temperature of the water to be filtered. Some filter cartridges are manufactured from a white cellulose fiber, some are manufactured of cotton wound fiber, and some filter cartridges are filled with activated carbon, which is also used for filtering and removing odor and other contaminants from the water. Different micron-size cartridges can be purchased for different degrees of filtering. Micron (or micrometer or μm) is a unit of measurement equal to one-thousandth of a millimeter. It is sometimes confusing to hear that 5 microns is larger that 10 microns, but the higher the micron number, the smaller the particle size that can be filtered.

Dirty water is cleaned of suspended solids as it works itself through the small porous openings in the filter media and onto the center opening in the filter cartridge. The cartridges eventually become so saturated with suspended solids that the water flow through the cartridge becomes diminished to the point of no flow and the cartridge filters have to be removed from the filter and thrown away. This type of filter works very well for systems where there is not a great deal of suspended matter in the water. For higher amounts of suspended matter the cost of replacing filter cartridges becomes prohibitive.

When a new cartridge filter is installed, the pressure drop across the filter is small. This means that the incoming pressure less the outgoing pressure is the pressure drop across the filter. As the filter is used the dissolved solids begin to fill up the void spaces in the filter cartridge and it becomes more difficult for the water to pass through the filter. This in turn increases the differential pressure drop from the inlet side of the filter to the outlet side of the filter. It is always good practice to install pressure gages on each side of any filter so that the pressure drop across the filter can be determined. This gives the operator a guide as to when to replace or backwash the filter.

Each cartridge filter is rated to pass a predetermined number of gallons per minute. Pressure and temperature also determines the flow rate in gal/min of a cartridge filter as well as the type of influent material passing through the filter. If one cartridge were designed to pass 6 gal/min through it and the system required 12 gal/min, it would require two cartridges to operate the system.

Sand Filters

Sand filters have an advantage over cartridge filters in that sand filters can be backwashed and then used again. Sand filters can also be automated to backwash automatically and then be returned to service. Sand filters can also handle larger volumes of dissolved solids before the filter bed becomes exhausted. Sand filters are best suited for keeping tower water clean.

The rate (gallons per minute) of a vertical pressure sand filter is determined by the surface area of the filter. Water is usually passed downward through different layers of finely divided filtering media, such as sand, quartz, and/or gravel. Sometimes the introduction of coagulants, a material that attaches itself to dissolved solids and then forms a larger solid, like jelly, is added. This larger particle makes it easier to remove solids in the water as it passes into the filter, because the solids are larger and it aids materially in the operation of a rapid rate of filtration. Common coagulants are alum, sulfate of aluminum, and/or iron sulfates.

Suitable piping connections are provided in the filter so that the filter may be backwashed by introducing water, under pressure, below the bottom strainer plate of the filter, forcing the water upward through the filter media bed, lifting the media, and removing the entrapped debris, then washing them away to drain.

By forcing water up through the media, the sand and other media are lifted and rearranged according to the specific gravity of each grain of media. This is the reason for the space that exists above the sand level in a sand filter. During the backwashing, the filter bed is expanded from 150 to over 200 percent of its original volume. The grains of sand, while suspended, rub against each other and aid in effecting proper cleaning. An excessive backwash rate can cause the expansion of mineral in the tank to be too great and the sand grains can wash out of the filter with the suspended solids. When the backwash water is cut off, there will no longer be any water pressure holding the media in suspension and the media will fall, redistributing itself evenly in the bed of the filter. The lighter, finer particles of media will form on top and the heavier media on the bottom.

For good backwash clean water should be used. This would require a clean filtered water holding tank downstream of the filter, so that water from the holding tank can be used for backwash. Automatic backwash control timers or differential pressure switches can be used to activate backwash.

Solid Separator

Solid separators actually separate the suspended solids from the water by centrifugal force, which is sensitive to particle weight and buoyancy and removes these solids through blowdown, while filters use a filter media, which is sensitive to micron size.

Because of this basic difference in operation between filters and separators, consideration when using separators should be given to partial weight and buoyancy rather than micron size. For instance, cottonwood seeds that float in the air are not easily removed with a separator, but easily removed in a filter. This is because cottonseed is very light and buoyant.

A separator generates a powerful centrifugal force by injecting water into the separator on a tangent at the top of the separator. This water when entering the separator is traveling at a very high velocity, but upon entering the separator the water has farther to travel, because of the larger circumference of the separator, and the velocity slows down.

Because the solids in the water are heavier than the water, the suspended solids are thrown against the outer walls of the separator and begin to slide down the outer walls, separating themselves from the water. The clean water in the separator now sees a low-pressure point at the top center of the separator and begins to pick up speed as it rushes toward the low-pressure point where the clean water leaves the separator.

As the solids slide down the outer walls of the separator, they enter a quiescent zone and quickly settle to the bottom of the separator. From time to time a valve in the bottom of the separator is opened and the pressure in the separator blows the collected suspended solids out to drain.

A simple field test is to see what kind of performance is expected from a centrifugal separator when pouring a representative sample of the dirty water to be processed, complete with solids, into a clear glass container and then shaking the glass until the solids are in suspension. After setting the glass down you can see what a centrifugal separator will do. A centrifugal separator will do a good job of removing the solids that settle to the bottom of the glass in the first 30 to 60 sec. With this test you can decide with confidence whether your requirements are best handled with a simple sidestream separator or with a more complex costly and maintenance-prone media-type filter.

Some advantages of a centrifugal separator over a media filter are as follows:

1.   A centrifugal separator uses very little water when dumping the suspended solids, whereas a media filter uses gallons of water for backwash, which then goes to the sewer along with chemicals used for treating the tower water.

2.   The cost of a centrifugal separator is much less than a media filter.

3.   The operating cost of a centrifugal separator is much less than a media filter.

4.   The space required for a centrifugal separator is much less than the space required for a media filter.

CHAPTER 12

GADGETS

There is one born every minute. Gadgets are devices that are sold by slick high-pressure salespeople to unsuspecting plant operators. They tell you that they can control scale, corrosion, and biological deposits in cooling systems and hydronic systems without using chemicals (See Fig. 12.1).

To date, none of the currently available group of devices or gadgets that operate on *electromagnetic, electrostatic, magnetic, sonic,* and other physical principles for the prevention of scale and corrosion in water has been proven to work. They have proven to be ineffective in unbiased tests conducted by qualified water corrosion research scientists in laboratories of major universities or independent research institutes such as The Massachusetts Institute of Technology or International Telephone and Telegraph (ITT).

Do not be in a rush to take the word of one of these high-pressure salespeople. They can make it sound as if they have the answer to all of your problems. Wait until a company you know, which is using gadgets, is ready for an *annual* inspection of their system and try to be on hand for the inspection. Your local boiler inspector should also be able to inform you of inspections he or she has made on equipment of companies that use gadgets.

If you have to try one of these gadgets, install it on a small, noncritical system, such as an air conditioner. Try not to be the first in your area or industry to try the device. Ask the salesperson to return when

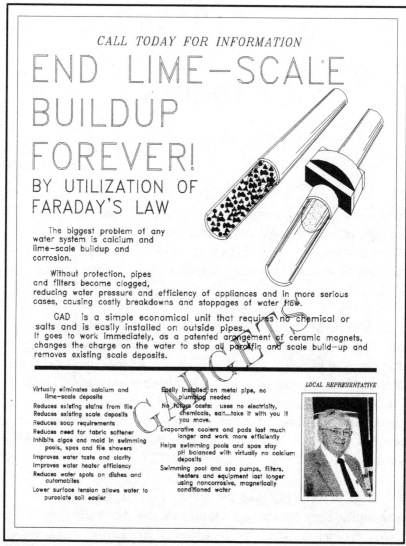

FIGURE 12.1 Gadget example.

possible to demonstrate a successful installation nearby and on equipment resembling your own. Remember, if a manager or an operator buys a gadget and has it installed it is probably going to cost a lot of up-front money. If the manager spends the company money and the device does not work, he or she will never want anyone to know that it did not work or that a mistake has been made and will probably tell you that it works great. Check all references carefully.

Stop and think! If these devices are so good, why don't the big water treatment companies sell them?

There are several good articles and reports regarding gadgets, sometimes called, *Nonchemical Devices* (NCD). It would be good to read them before embarking on an expensive NCD purchase and installation. Some are available from the following sources.

> Nonchemical Devices for Water Treatment
>> by Paul Puckorius
>> Puckorius & Associates Inc.
>> P.O. Box 2440
>> Evergreen, Colorado 80439

> National Association of Corrosion Engineers (NACE)
>> P.O. Box 218340
>> Houston, Texas 77218

One of the latest papers regarding NCD was written by NACE . The following is a portion of their Corrosion 95 Minutes, taken on March 29,1995 from Unit Committee T-7 K Nonchemical Water Treating Devices.

As can be seen by reading only a short portion of this report, the airline facility has sustained considerable corrosion and biological growth problems in their system because of the use of magnets. No consideration was given with regard to the cost of installing the magnets. The airline facility has now reverted back to the use of chemicals to protect its equipment.

RECENT REPORT

Consumers Union recently tested a magnetic water treatment device sized for household use (see the February 1996 issue of <u>Consumer Reports</u>). Technicians installed two water heaters side by side and put

the magnet on one. The heaters were operated for two years on a
water containing 200 ppm calcium hardness. After each heater had

---

4.2   Site visits by Helmut Brutman and Craig Schinzer. The
        first site visit was made to an airline facility. This facility
        chose magnets in an attempt to be "environmentally
        friendly." The facility uses them for all of their systems,
        including boilers and cooling towers. The boiler also has
        a deaerating feedwater heater and softener. Inspection
        showed corrosion problems are evident already in both the
        boilers and condensate return system. The boiler also has
        a deaerating feedwater heater and softener. Inspection
        showed corrosion problems are evident already in both the
        boilers and condensate return system. The cooling towers
        have an accumulation of sand-like deposits and significant
        biological growth problems. The facility noted that the
        on-line conductivity probe and laboratory conductivity
        probe consistently showed different readings.   Some
        investigation is needed to determine if this is real or an
        artifact of calibration differences. Consideration is now
        being given to using sulfite for oxygen scavenging and
        amine for condensate return protection. They are also
        contemplating oxidizing and nonoxidizing biocides for
        biological control in the cooling system.

FIGURE 12.2   Copy of report.

heated more than 10,000 gallons of water, the heaters were taken
apart. The water heater with the magnetic device contained scale of
the same quantity and texture as the untreated heater. In this test the
device did not provide any measurable benefit.

Another source for information on gadgets would be your water
treatment supplier or the city boiler inspector.

CASE HISTORIES

The following is a list of some well-known companies that have tried using the magnets on steam boilers, hydronic systems, cooling towers, etc., and have now gone back to conventional water treatment.

Martin Marietta Corporation

This firm installed a new Cleaver Brooks Model 4 steam boiler using magnets on the incoming feedwater system to the boiler. Within six months the company had to clean the boiler tubes twice and replace several of the fire tubes in the boiler, which were plugged solid with scale.

Wester Steel & Boiler Co.
Denver, Colorado

Bronco Football Training Facility

They were talked into using magnets on their domestic water system to prevent scale. After scaling up their cooling towers, they reverted back to conventional water treatment.

Sage Industries, Inc.
Denver, Colorado

Denver International Airport

A large new airport was built in Denver, Colorado, with the specification that magnets be installed on all Heating, Ventilating, and Air-Conditioning (HVAC) piping to the main terminal. Before the airport was officially opened, airport management had the magnets removed, at a great expense to the city of Denver, because management knew that they would not do the job.

Aqua-Chem, Inc.
Denver, Colorado

Monfort Meat Packing

This company installed magnets in a one-pass cooling tower water system. The towers scaled up within one month, and the magnets were removed.

The Nunn Co. Inc.
Greeley, Colorado

APPENDIX A

CONVERSION FACTORS

To convert	Multiply by	To obtain

A

To convert	Multiply by	To obtain
acre-feet	4.356×10^4	cubic feet
acre-feet	3.259×10^5	gallons

B

To convert	Multiply by	To obtain
barrels (U.S. dry)	7.056×10^3	cubic inches
barrels (U.S. liquid)	3.15×10^1	gallons
Btu	7.7816×10^2	ft. lbs
Btu	3.927×10^{-4}	horsepower hours
Btu	2.928×10^{-4}	kilowatt hours
Btu/hr	2.162×10^{-1}	ft. lbs / sec
Btu/hr	3.929×10^{-4}	horsepower
Btu/hr	2.931×10^{-1}	watts
Btu/min	1.296×10^1	ft lbs/sec
Btu/min	2.356×10^{-2}	horsepower
Btu/min	1.757×10^{-2}	kilowatts
Btu/min	1.757×10^1	watts
Btu/ft²/min	1.22×10^{-1}	watts/square in

C

To convert	Multiply by	To obtain
centimeters	3.281×10^{-2}	feet
centimeters	3.937×10^{-1}	inches
centimeters	$1. \times 10^{-5}$	kilometers
centimeters	$1. \times 10^{-2}$	meters
centimeters	$1. \times 10^1$	millimeters
centimeters	3.937×10^2	mils
centimeters	1.094×10^{-2}	yards
centimeters	$1. \times 10^4$	microns
centimeters of mercury	$1.316 \cdot 10^{-2}$	atmospheres
centimeters of mercury	4.461×10^{-1}	ft of water
centimeters of mercury	2.785×10^1	lbs / ft³
centimeter of mercury	1.934×10^{-1}	lbss / in²
cm / s	1.969	ft / min
cm / s	3.281×10^{-2}	ft / s
cm / s / s	3.281×10^{-2}	ft / s / s
cubic cm	3.531×10^{-5}	cubic ft
cubic cm	6.102×10^{-2}	cubic inch
cubic cm	2.642×10^{-4}	gal. (U.S.liq.)
cubic cm	1.0×10^{-3}	liters
cubic cm	2.113×10^{-3}	pints (U.S. liq)
cubic cm	1.057×10^{-3}	qt (U.S. liq)
cubic feet	2.8320×10^4	cu cms
cubic feet	1.728×10^3	cu inches
cubic feet	2.832×10^{-2}	cu meters
cubic feet	3.704×10^{-2}	cu yards
cubic feet	7.48052	gal (U.S. liq)
cubic feet	2.832×10^1	liters.
cubic feet	5.984×10^1	pints (U.S. liq)
cubic feet	2.992×10^1	qt (U.S. liq)
cubic ft/min	4.72×10^2	cu cms./ s
cubic ft/min	1.247×10^{-1}	gallons / s
cubic ft/min	4.720×10^{-1}	liters / s
cubic ft/min	6.243×10^1	lb water / min
cubic ft/s	6.4631×10^{-1}	million gal / day
cubic ft/s	4.4883×10^2	gal / min
cubic inches	1.639×10^1	cu cms
cubic inches	5.787×10^{-4}	cu ft
cubic inches	1.639×10^{-5}	cu meters
cubic inches	2.143×10^{-5}	cu yards
cubic inches	4.329×10^{-3}	gallons
cubic inches	1.639×10^{-2}	liters
cubic inches	3.463×10^{-2}	pt (U.S. liq)
cubic inches	1.732×10^{-2}	qt (U.S. liq)
cubic meters	1.0×10^6	cu cms
cubic meters	3.531×10^1	cu ft
cubic meters	6.1023×10^4	cu inches
cubic meters	1.308	cu yards
cubic meters	2.642×10^2	gal (U.S. liq)
cubic meters	1.0×10^3	liters
cubic meters	2.113×10^3	pt (U.S. liq)
cubic meters	1.057×10^3	qts (U.S. liq)
cubic yards	7.646×10^5	cu. cms
cubic yards	2.7×10^1	cu ft
cubic yards	4.6656×10^4	cu inches
cubic yards	7.646×10^{-1}	cu meters
cubic yards	2.02×10^2	gal (U.S. liq)
cubic yards	7.646×10^2	liters
cubic yards	1.6159×10^3	pt (U.S. liq)
cubic yards	8.079×10^2	qt (U.S. liq)
cub yd / min	4.5×10^{-1}	cubic ft / s
cub yd / min	3.367	gal / s
cub yd / min	1.274×10^1	liters/s

APPENDIX A

To convert	Multiply by	To obtain
D		
days	8.64×10^4	seconds
days	1.44×10^3	minutes
days	2.4×10^1	hours
E		
F		
feet	3.048×10^1	centimeters
feet	3.048×10^{-4}	kilometers
feet	3.048×10^{-1}	meters
feet	3.048×10^2	millimeters
feet	1.2×10^4	mils
feet of water	2.95×10^{-2}	atmospheres
feet of water	8.826×10^{-1}	in of mercury
feet of water	3.048×10^{-2}	kg / sq cm
feet of water	6.243×10^1	pounds / sq ft
feet of water	4.335×10^{-1}	pounds / sq in
feet / min	5.080×10^{-1}	cms / s
feet / min	1.667×10^{-2}	feet / s
feet / min	1.829×10^{-2}	km / hr
feet / min	3.048×10^{-1}	meters / min
feet / min	1.136×10^{-2}	miles / hr
feet / s	3.048×10^1	cm / s
feet / s	1.097	kms / hr
feet / s	1.829×10^1	meters / min
feet / s	6.818×10^{-1}	miles / hr
feet / s	1.136×10^{-2}	miles / min
feet / s / s	3.048×10^1	cm / s / s
feet / s / s	1.097	km / hr / s
feet / s / s	3.048×10^{-1}	meters / s / s
feet /s / s	6.818×10^{-1}	miles / hr / s
foot-pounds	1.286×10^{-3}	Btu
foot-pounds	5.050×10^{-7}	horsepower hr
foot-pounds	3.766×10^{-7}	kilowatt hrs
ft lb / min	1.286×10^{-3}	Btu / min
ft lb / min	1.667×10^{-2}	foot-pounds / s
ft lb / min	3.030×10^{-5}	horsepower
ft lb / min	3.241×10^{-4}	kg calories / min
ft lb / min	2.260×10^{-5}	kilowatts
ft lb / s	4.6263	Btu / hr
ft lb / s	7.717×10^{-2}	Btu / min
ft lb / s	1.818×10^{-3}	horsepower
ft lb / s	1.945×10^{-2}	kg calories / min
ft lb / s	1.356×10^{-3}	kilowatts

To convert	Multiply by	To obtain
G		
gallons	3.785×10^3	cu cm
gallons	1.337×10^{-1}	cu feet
gallons	2.31×10^2	cu inches
gallons	3.785×10^{-3}	cu meters
gallons	4.951×10^{-3}	cu yards
gallons	3.785	liters
gallons (U.S.)	8.3267×10^{-1}	gallons (imp)
gal of water	8.337	lb of water
gal / min	2.228×10^{-3}	cu feet / s
gall./ min	6.308×10^{-2}	liters / s
gall / min	8.0208	cu feet / hr
gr / U.S. gal	1.7118×10^1	ppm
gr / U.S. gal	1.4286×10^2	lb/ million gal
grams	1.0×10^{-3}	kilograms
grams	1.0×10^3	milligrams
grams	2.205×10^{-3}	pounds
grams / cm	5.6×10^{-3}	pounds / in
grams / cu cm	6.243×10^1	pounds / cu ft
grams / cu cm	3.613×10^{-2}	pounds / cu in
grams / cu cm	3.405×10^{-7}	lbs / million ft
grams / liter	5.8417×10^1	grains / gal
grams / liter	8.345	lb /1000 gal
grams / liter	6.2427×10^{-2}	lb / cu ft
grams / sq cm	2.0481	lb / sq ft
gram-calories	3.9683×10^{-3}	Btu
gram-calories	3.086	ft lb
gram-calories	1.5596×10^{-6}	hp hours
gram-calories	1.162×10^{-6}	kilowatt hrs
gram-calories	1.162×10^{-3}	watt hrs
gram cal / s	1.4286×10^1	Btu / hr
gram c-meter	9.297×10^{-8}	Btu
gram-c-meter	2.343×10^{-8}	kg calories
gram-c-meter	1.0×10^{-5}	kg meters
H		
hectares	1.076×10^5	sq feet
hectograms	1.0×10^2	grams
hectoliters	1.0×10^2	liters
hectometers	1.0×10^2	meters
hectowatts	1.0×10^2	watts
horsepower	4.244×10^1	Btu / min
horsepower	3.3×10^4	foot lbs / min
horsepower	5.5×10^2	foot lbs / sec
horsepower	9.863×10^{-1}	h power
horsepower	1.014	h power (metric)

APPENDIX A

To convert	Multiply by	To obtain
horsepower	1.068×10^1	kg calories / min
horsepower	7.457×10^{-1}	kilowatts
horsepower	7.457×10^2	watts
h.p (boiler)	3.352×10^4	Btu / hr
h.p (boiler)	9.803	kilowatts
h p h	2.547×10^3	Btu
h p h	1.98×10^6	ft lb
hp h	6.4119×10^5	gram calories
hp h	7.457×10^{-1}	kilowatt hrs
hours	4.167×10^{-2}	days
hours	5.952×10^{-3}	weeks
hours	3.6×10^3	seconds

I

To convert	Multiply by	To obtain
inches	2.540	centimeters
inches	2.54×10^{-2}	meters
inches	1.578×10^{-5}	miles
inches	2.54×10^1	millimeters
inches	1.0×10^3	mils
inches	2.778×10^{-2}	yards
inch mercury	3.342×10^{-2}	atmospheres
inch mercury	1.133	ft of water
inch mercury	3.453×10^{-2}	kg / sq cm
inch mercury	3.453×10^2	kg / sq meter
inch-mercury	7.073×10^1	pounds / sq ft
inch mercury	4.912×10^{-1}	pounds / sq in
in water (4°C)	2.458×10^{-3}	atmospheres
in water (4°C)	7.355×10^{-2}	inches of mercury
in water (4°C)	2.54×10^{-3}	kg / sq cm
in water (4°C)	5.781×10^{-1}	ounces / sq in
in water (4°C)	5.204	lb / sq ft
in water (4°C)	3.613×10^{-2}	lb / sq in

J

To convert	Multiply by	To obtain
joules	7.736×10^{-1}	ft lb
joules	2.389×10^{-4}	kg calorie
joules	1.020×10^{-1}	kg meter
joules	2.778×10^{-4}	watt hrs

K

To convert	Multiply by	To obtain
kilograms	9.8066×10^5	dynes
kilograms	1.0×10^3	grams
kilograms	2.2046	pounds

To convert	Multiply by	To obtain
kilograms	1.102×10^{-3}	tons (short)
kilograms	3.5274×10^1	ounces avdp
kg / cu m	1.0×10^{-3}	grams / cu cm
kg / cu m	6.243×10^{-2}	lb / cu ft
kg / cu m	3.613×10^{-5}	lb / cu in
kg / cu m	3.405×10^{-10}	lb./ mil foot
kg / m	6.72×10^{-1}	lb / ft
kg / sq cm	9.678×10^{-1}	atmospheres
kg / sq m	3.281×10^{-3}	feet of water
kg / sq cm	2.896×10^1	inches of mercury
kg / sq cm	1.422×10^1	pounds / sq in
kg /sq m	9.678×10^{-5}	atmospheres
kg cal	3.968	Btu
kg cal	3.086×10^3	ft lb
kg cal	4.183×10^3	joules
kg cal	4.269×10^2	kg meters
kg cal	4.163×10^{-3}	kilowatt hrs
kg cal / m	5.143×10^1	ft lb./ s
kg cal / m	9.351×10^{-2}	horsepower
kg cal / m	6.972×10^{-2}	kilowatts
kg meters	9.296×10^{-3}	Btu
kg meters	7.233	ft lb
kg meters	9.807	joules
kg meters	2.342×10^{-3}	kg calories
kg meters	2.723×10^{-6}	kilowatt hrs
kiloliters	1.0×10^3	liters
kiloliters	1.308	cubic yards
kiloliters	3.5316×10^1	cubic feet
kiloliters	2.6418×10^2	gal (U.S. liq)
kilometers	1.0×10^5	centimeters
kilometers	3.281×10^3	feet
kilometers	3.937×10^4	inches
kilometers	1.0×10^3	meters
kilometers	6.214×10^{-1}	miles (statute)
kilometers	1.0×10^6	millimeters
kilometers	1.0936×10^3	yards
k meters / hr	2.778×10^1	cm / s
k meters / hr	5.468×10^1	feet / min
k meters / hr	9.113×10^{-1}	feet / s
k meters / hr	1.667×10^1	meters / min
k meters / hr	6.214×10^{-1}	miles / hr
k m / hr./s	2.778×10^1	cm sec / s
k m / hr / s	9.113×10^{-1}	ft / sec / s
k m / hr / s	2.778×10^{-1}	meters / s / s
k m / hr / s	6.214×10^{-1}	miles / hr / s
kilowatts	5.692×10^1	Btu / min
kilowatts	4.426×10^4	ft lb / min
kilowatts	7.376×10^2	ft lb / sec
kilowatts	1.341	horsepower
kilowatts	1.434×10^1	kg calorie / min
kilowatts	1.0×10^3	watts
kilowatt hrs	3.413×10^3	Btu
kilowatt hrs	2.655×10^{-6}	ft lb

APPENDIX A

To convert	Multiply by	To obtain
kilowatt hrs	8.5985×10^5	gram calories
kilowatt hrs	1.341	horsepower hours
kilowatt hrs	3.6×10^6	joules
kilowatt hrs	8.605×10^2	kg calories
kilowatt hrs	3.53	lb of water evap. from and at 212^0 F
kilowatt hrs	2.275×10^1	lb of water raised from 62° F to 212^0 F

L

To convert	Multiply by	To obtain
liters	1.0×10^3	cu cm
liters	3.531×10^{-2}	cu ft
liters	6.102×10^1	cu inches
liters	1.0×10^{-3}	cu meters
liters	1.308×10^{-3}	cu yards
liters	2.642×10^{-1}	gal (U.S. liq)
liters	2.113	pints (U.S. liq)
liters	1.057	quart (U.S. liq)
liters / min	5.886×10^{-4}	cu ft / s
liters / min	4.403×10^{-3}	gals / s

M

To convert	Multiply by	To obtain
meters	1.0×10^2	centimeters
meters	3.281	feet
meters	3.937×10^1	inches
meters	1.0×10^{-3}	kilometers
meters	6.214×10^{-4}	miles (statute)
meters	1.0×10^3	millimeters
meters	1.094	yards
meters / min	1.667	cm / s
meters / min	3.281	ft / min
meters / min	5.468×10^{-2}	ft / s
meters / min	6.0×10^{-2}	km hr
meters / min	3.728×10^{-2}	miles / hr
meters / s	1.968×10^2	feet / min
meters / s	3.281	ft / s
meters / s	3.6	kilometers / hr
meters / s	6.0×10^{-2}	kilometers / min
meters / s	2.237	miles / hr
meters / s	3.728×10^{-2}	miles / min
m / s./ s	1.0×10^2	cms./ s / s
m / s./ s	3.281	ft / s / s

To convert	Multiply by	To obtain
m / s / s	3.6	km / hr / s
m / s./ s	2.237	miles / hr / s
m kilograms	1.0×10^5	cm grams
m kilograms	7.233	ft pound
miles (statute)	1.609×10^5	centimeters
miles (statute)	5.280×10^3	feet
miles (statute)	6.336×10^4	inches
miles (statute)	1.609	kilometers
miles (statute)	1.609×10^3	meters
miles (statute)	1.760×10^{-3}	yards
miles / hr	4.470×10^1	cm / s
miles / hr	8.8×10^1	ft / min
miles / hr	1.467	ft / s
miles / hr	1.6093	km / hr
miles / hr	2.682×10^{-2}	km / min
miles / hr	2.682×10^1	meters / min
miles / hr	1.667×10^{-2}	miles / min
miles / hr/s	4.47×10^1	cm / s / s
miles/ hr/s	1.467	ft / s / s
miles/ hr/s	1.6093	km / hr / s
miles/ hr/s	4.47×10^{-1}	meters / s/ s
miles/ min	2.682×10^3	cm / s
miles/ min	8.8×10^1	feet / s
miles/ min	1.6093	km / min
miles/ min	6.0×10^1	miles / hr
milligrams	1.5432×10^{-2}	grains
milligrams	1.0×10^{-3}	grams
m grams/liter	1.0	parts / million
milliliters	1.0×10^{-3}	liters
millimeters	1.0×10^{-1}	centimeters
millimeters	3.281×10^{-3}	feet
millimeters	3.937×10^{-2}	inches
millimeters	1.0×10^{-6}	kilometers
millimeters	1.0×10^{-3}	meters
millimeters	6.214×10^{-7}	miles
millimeters	3.937×10^1	mils
millimeters	1.094×10^{-3}	yards
million g/day	1.54723	cu ft / s
mils	2.54×10^{-3}	centimeters
mils	8.333×10^{-5}	feet
mils	1.0×10^{-3}	inches
mils	2.54×10^{-8}	kilometers
mils	2.778×10^{-5}	yards
min (time)	9.9206×10^{-5}	weeks
min (time)	6.944×10^{-4}	days
min (time)	1.667×10^{-2}	hours
min (time)	6.0×10^1	seconds

To convert	Multiply by	To obtain
N		
O		
ounces	8.0	drams
ounces	4.375×10^2	grains
ounces	2.8349×10^1	grams
ounces	6.25×10^{-2}	pounds
ounces	3.125×10^{-5}	tons (short)
ounces (fluid)	1.805	cu inch
ounces (fluid)	2.957×10^{-2}	liters
ounces / sq in	4.309×10^3	dynes / sq cm
ounces / sq in	6.25×10^{-2}	lb /sq in
ounces	7.8123×10^{-3}	gal
P		
parts / mil	5.84×10^{-2}	grains / U.S. gal
parts / mil	8.345	lb / million gal
pints (liquid)	4.732×10^2	cubic cm
pints (liquid)	1.671×10^{-2}	cubic ft
pints (liquid)	2.887×10^1	cubic inches
pints (liquid)	4.732×10^{-4}	cubic meters
pints (liquid)	6.189×10^{-4}	cubic yards
pints (liquid)	1.25×10^{-1}	gallons
pints (liquid)	4.732×10^{-1}	liters
pints (liquid)	5.0×10^{-1}	quarts (liquid)
pounds	2.56×10^2	drams
pounds	4.448×10^5	dynes
pounds	7.0×10^3	grains
pounds	4.5359×10^2	grams
pounds	4.536×10^{-1}	kilograms
pounds	1.6×10^{-1}	ounces
pounds water	1.602×10^{-2}	cu ft
pounds water	2.768×10^1	cu inches
pounds water	1.198×10^{-1}	gallons
lb water / min	2.670×10^{-4}	cu ft./ s
pound ft	1.3825×10^4	cm grams
pound ft	1.383×10^{-1}	meter kg
lb / cu ft	1.602×10^{-2}	grams / cu cm
lbs / cu ft	1.602×10^1	kg / cu meter
lb / cu ft	5.787×10^{-4}	pounds / cu inch
lb / cu ft	5.456×10^{-9}	pounds / mil ft
lb / cu in	2.768×10^1	grams / cu cm
lb / cu in	2.768×10^4	kg / cu meter
lb / cu in	1.728×10^3	pounds / cu ft
lb / cu in	9.425×10^{-6}	pounds / mil ft
lb / foot	1.488	kg / meter

To convert	Multiply by	To obtain
lb / in	1.786×10^2	grams / cm
lb / mil ft	2.306×10^6	grams / cu cm
lb / sq ft	4.725×10^{-4}	atmospheres
lb / sq ft	1.602×10^{-2}	feet of water
lb / sq ft	1.414×10^{-2}	inches of mercury
lb / sq ft	4.882	kg / sq meter
lb / sq ft	6.944×10^{-3}	pounds / sq in
lb / sq in	6.804×10^{-2}	atmospheres
lb / sq in	2.307	ft of water
lb / sq in	2.036	inches of mercury
lb / sq in	7.031×10^2	kg / sq meter
Q		
qt (liquid)	9.464×10^2	cu cms
qt (liquid)	3.342×10^{-2}	cu ft
qt (liquid)	5.775×10^1	cu inches
qt (liquid)	9.464×10^{-4}	cu meters
qt (liquid)	1.238×10^{-3}	cu yards
qt (liquid)	2.5×10^{-1}	gallons
qt (liquid)	9.463×10^{-1}	liters
R		
S		
square cm	1.973×10^5	circular mils
square cm	1.076×10^{-3}	sq ft
square cm	1.550×10^{-1}	sq inches
square cm	1.0×10^{-4}	sq meters
square ft	1.833×10^8	circular mils
square ft	9.29×10^2	sq cm
square ft	1.44×10^2	sq inches
square ft	9.29×10^{-2}	sq meters
square ft	1.111×10^{-1}	sq yards
square inch	6.452	sq cm
square inch	6.944×10^{-3}	sq ft
square inch	6.452×10^2	sq millimeters
square inch	1.0×10^6	sq mils
square inch	7.716×10^{-4}	sq yards
square meters	1.0×10^4	sq cm
square meters	1.076×10^1	sq ft
square meters	1.55×10^3	sq inches
square meters	1.0×10^6	sq millimeters
square meters	1.196	sq yards
sq millimeters	1.0×10^{-2}	sq cm
sq millimeters	1.076×10^{-5}	sq ft
sq millimeters	1.55×10^{-3}	sq inches

To convert	Multiply by	To obtain
square mils	6.452×10^{-6}	sq cm
square mils	1.0×10^{-6}	sq inches

T

To convert	Multiply by	To obtain
temp (^{O}C) +273		
1.0		absolute (^{O}K)
temp (^{O}C) +17.78		
1.8		temperature (^{O}F)
temp (^{O}F) +460		
1.0		absolute temp (R)
temp (^{O}F) -32		
5/9		temperature (^{O}C)
tons (long)	1.016×10^{3}	kilograms
tons (long)	2.24×10^{3}	pounds
tons (long)	1.12	tons (short)
tons (metric)	1.0×10^{3}	kilograms
tons (metric)	2.205×10^{3}	pounds
tons (short)	9.0718×10^{-2}	kilograms
tons (short)	3.2×10^{4}	ounces
tons (short)	2.0×10^{3}	pounds
tons (short)	9.0718×10^{2}	kilograms
tons (short)	3.2×10^{4}	ounces
tons (short)	2.9166×10^{4}	ounces (troy)
tons (short)	2.0×10^{3}	pounds
tons (short)	2.43×10^{3}	pounds (troy)
tons (short)	8.929×10^{-1}	tons (long)
tons (short)	9.078×10^{-1}	tons (metric)
tons (s) / sq ft	9.765×10^{3}	kg / sq meter
tons (s) / sq ft	1.389×10^{1}	pounds / sq in
tons (s) / sq in	1.406×10^{6}	kg / sq meter
tons (s) / sq in	2.0×10^{3}	pounds / sq in
tons of water / 24 hr	8.333×10^{1}	lb of water / hr
tons of water / 24 hr	1.6643×10^{-1}	gallons / min
tons of water / 24 hr	1.3349	cu. ft / hr

U

V

W

To convert	Multiply by	To obtain
watts	3.4129	Btu / hr
watts	5.688×10^{-2}	Btu / min
watts	4.427×10^{1}	ft lbs / min
watts	7.378×10^{-1}	ft lbs / sec
watts	1.0×10^{-3}	kilowatts
watt hours	3.413	Btu
watt hours	2.656×10^{3}	ft lb
watt hours	1.341×10^{-3}	hp hours
watt hours	8.605×10^{2}	gram calories
watt hours	8.605×10^{-1}	kilogram calories
watt hours	3.672×10^{2}	kilogram meters
watt hours	1.0×10^{-3}	kilowatt hours
weeks	1.68×10^{2}	hours
weeks	1.008×10^{4}	minutes
weeks	6.048×10^{5}	seconds

X

Y

Z

APPENDIX B

USEFUL PHYSICAL CONSTANTS

1 boiler hp = 10 ft² heating surface

1 boiler hp = 34.5 lb water evaporated per hour

1 boiler hp = 4 gal water evaporated per hour

1 U.S. gal of water = 231 in³

1 U.S. gal of water = 8.33 lb

1 ft³ water = 7.48 gal of water

1 ft³ water = 62.3 lb

head (in ft) = psi • 2.13 ÷ sg

psi = head (in ft) • sg ÷ 2.13

1 ft of head (water) = 0.434 lb/in² (water)

1 grain per gallon = 17.1 parts per million

grains per U.S. gal = pounds per 1000 gallons • 7

1 lb = 7000 grains

parts per million = pounds per 1000 gal • 120

gallons per 24 hours = gallons per minute • 1440

gallons per minute = gallons per 24 hours ÷ 1440

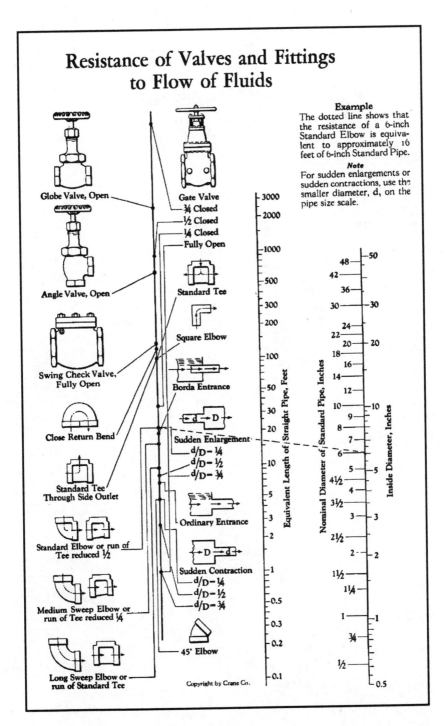

Resistance of Valves and Fittings to Flow of Fluids

Example

The dotted line shows that the resistance of a 6-inch Standard Elbow is equivalent to approximately 16 feet of 6-inch Standard Pipe.

Note

For sudden enlargements or sudden contractions, use the smaller diameter, d, on the pipe size scale.

Globe Valve, Open

Angle Valve, Open

Swing Check Valve, Fully Open

Close Return Bend

Standard Tee Through Side Outlet

Standard Elbow or run of Tee reduced ½

Medium Sweep Elbow or run of Tee reduced ¼

Long Sweep Elbow or run of Standard Tee

Gate Valve
¾ Closed
½ Closed
¼ Closed
Fully Open

Standard Tee

Square Elbow

Borda Entrance

Sudden Enlargement
d/D – ¼
d/D – ½
d/D – ¾

Ordinary Entrance

Sudden Contraction
d/D – ¼
d/D – ½
d/D – ¾

45° Elbow

3000
2000
1000
500
300
200
100
50
30
20
10
5
3
2
1
0.5
0.3
0.2
0.1

Equivalent Length of Straight Pipe, Feet

Nominal Diameter of Standard Pipe, Inches

48
42
36
30
24
22
20
18
16
14
12
10
9
8
7
6
5
4½
4
3½
3
2½
2
1½
1¼
1
¾
½

Inside Diameter, Inches

50
30
20
10
5
3
2
1
0.5

Copyright by Crane Co.

Flow of Water Through Schedule 40 Steel Pipe

Pressure Drop per. 100 feet and Velocity in Schedule 40 Pipe for Water at 60 F.

Discharge Gallons per Minute	Velocity Feet per Second	Press. Drop Lbs. per Sq. In.	Velocity Feet per Second	Press. Drop Lbs. per Sq. In.	Velocity Feet per Second	Press. Drop Lbs. per Sq. In.	Velocity Feet per Second	Press. Drop Lbs. per Sq. In.	Velocity Feet per Second	Press. Drop Lbs. per Sq. In.
	1/2"		**3/4"**		**1"**		**1 1/4"**			
1	1.06	0.6	0.602	0.155	0.371	0.048			**1 1/2"**	
2	2.11	2.1	1.2	0.526	0.743	0.164	0.429	0.044		
3	3.17	4.33	1.81	1.09	1.114	0.336	0.644	0.09	0.473	0.043
4	4.22	7.42	2.41	1.83	1.49	0.565	0.918	0.15	0.63	0.071
5	5.28	11.2	3.01	2.75	1.86	0.835	1.073	0.223	0.788	0.104
	2"									
6	0.574	0.044	3.61	3.84	2.23	1.17	1.29	0.309	0.946	0.145
8	0.765	0.073	**2 1/2"**		3.97	1.99	1.72	0.518	1.26	0.241
10	0.956	0.108			**3"**		2.15	0.774	1.58	0.361
15	1.43	0.224	1.01	0.094			3.22	1.63	2.37	0.755
20	1.91	0.375	1.34	0.158	0.868	0.056	4.29	2.78	3.16	1.28
25	2.39	0.561	1.68	0.234	1.09	0.083	**4"**		3.94	1.93
30	2.87	0.786	2.01	0.327	1.3	0.114			4.73	2.72
35	3.35	1.05	2.35	0.436	1.52	0.151	0.882	0.041	5.52	3.64
40	3.83	1.35	2.68	0.556	1.74	0.192	1.01	0.052	6.3	4.65
45	4.3	1.67	3.02	0.668	1.95	0.239	1.13	0.064	7	5.85
50	4.78	2.03	3.35	0.839	2.17	0.288	1.26	0.076	**5"**	
60	5.74	2.87	4.02	1.18	2.6	0.406	1.51	0.107		
70	6.7	3.84	4.69	1.59	3.04	0.54	1.76	0.143	1.12	0.047
80	7.65	4.97			3.47	0.687	2.02	0.18	1.28	0.06
90	8.6	6.2	**6"**		3.91	0.861	2.27	0.224	1.44	0.074
100	9.56	7.59	1.11	0.036	4.34	1.05	2.52	0.272	1.6	0.09
125	11.97	11.76	1.39	0.055	5.43	1.61	3.15	0.415	2.01	0.135
150	14.36	16.7	1.67	0.077	6.51	2.24	3.78	0.58	2.41	0.19
175	16.75	22.3	1.94	0.102			4.41	0.774	2.81	0.253
200	19.14	28.8	2.22	0.13	**8"**		5.04	0.985	3.21	0/323
225			2.5	0.162	1.44	0.043	5.67	1.23	3.61	0.401
250			2.78	0.195	1.6	0.051	6.3	1.46	4.01	0.495
275			3.05	0.234	1.76	0.061	6.93	1.79	4.41	0.583
300			3.33	0.275	1.92	0.072	7.56	2.11	4.81	0.683
325			3.61	0.32	2.08	0.083	8.19	2.47	5.21	0.797
350			3.89	0.367	2.24	0.095	8.82	2.84	5.62	0.919
375			4.16	0.416	2.4	0.108	9.45	3.25	6.02	1.05
400			4.44	0.471	2.56	0.121	10.08	3.68	6.42	1.19
425			4.72	0.529	2.73	0.136	10.71	4.12	6.82	1.33
450			5	0.59	2.89	0.151	11.34	4.6	7.22	1.48

INDEX

INDEX

INDEX

INDEX

INDEX

INDEX

INDEX

INDEX

INDEX